Sheshat Anthology
Volume 3

Sheshat Anthology
Volume 3

Contributing Authors:

Ai Ling Zhu | Amal Rizvi Ananda Majumdar | Ananna Arna
Andrew Clement | Angela Kazmierczak | Ashley Meelu
Austin Albert Mardon | Catherine Mardon | Chitrini Tandon
Daivat Bhavsar | Eman Zaheer | Gina Schopfer
Ismael Mohammedally | Janani Rajendra | Jasrita Singh
Jessica Jutras | Jilene Malbeuf |Lajendon Jeyakumar
Leah Sarah Peer | Lillian Zhang | Lina Lombo | Mehvish Masood
Muzammil Syed | Navneet Kang | Neha Saroya
Sriraam Sivachandran | Syed Muhammad Ali Salman
Tian Jao | Vivek Kannan | Viveka Pimenta | Yang Zhao
Zain Siddiqui

Editors:

Daivat Bhavsar | Catherine Mardon | Ethan Saldana

Typeset and Cover:

Ethan Saldana

GM
PRESS

2021

First Printing: 2021
ISBN 978-1-77369-221-0

Golden Meteorite Press
103 11919 82 St NW
Edmonton, AB T5B 2W3
www.goldenmeteoritepress.com

Table of Contents

Acknowledgements

The publication of the Sheshat Anthology Volume 3 is financially supported by the Antarctic Institute of Canada and TakingltGlobal charities.

The SHESHAT Volume 3 is an anthology assembled under the supervision of Drs. Austin and Catherine Mardon. This work will be published and promoted to different platforms such as Lulu, Google Scholar, and PubMed under the Antarctic Institute of Canada (AIC) Charity.

The SHESHAT Volume 3 is not a double-blind peer-reviewed journal as most journals; however, all articles are peer-reviewed thoroughly by experienced premedical and graduate students, and Dr. Mardon. The articles accepted in this paper are authored by skilled Writers of the Antarctic Institute of Canada. This anthology serves to appreciate and showcase youth scholarly research in the fields of gender studies, COVID-19, and socioeconomic aspects of daily living to name a few.

The AIC would like to acknowledge the #RisingYouth Grant offered by the TakingItGlobal Charity to many Article Writers to fund their project and publication. There are no conflicts of interests to declare.

Special Thanks to the Editor, Daivat Bhavsar, and the Graphic Designer Ethan Saldana, for their relentless efforts in assembling SHESHAT Volume 3.

THE CULTURAL SIGNIFICANCE OF THE EXISTENCE OF A METEORITE FOR THE CULT WORSHIP OF ARTEMIS IN EPHESUS.

Jilene Malbeuf[1] and Austin Mardon[2], [1]University of Alberta History and Classics department, Edmonton, Alberta, Canada, [2]Antarctic Institute of Canada.

Introduction:

The religious history of Ephesus (in modern day Turkey) is abnormally resistant to outside influences for its temporal and geographical setting, and this can be explained by the postulation that its strong devotion to Artemis is a result of their access to a meteorite as a cult statue. Establishing the presence of a meteorite as a part of the worship practices in Ephesus is problematic, as there is only one extant written record that suggests such a presence, in Acts 19:35, "is there anyone who does not know that the city of the Ephesians is temple keeper of the great Artemis, and of the image which fell down from Zeus?" The Greek term διοπετής translates to mean "descended from Zeus," and is a common description of meteorites which emphasizes a connection to the divine and the sacred status of that stone as a result of this connection. Meteorites create a sacred space upon impact with the earth, in the same way as a bolt of lightening would sanctify the space because Zeus himself has established a direct connection to it.

Temple of Artemis:

There are several instances in antiquity where a specific location remains sacred long after the precise reason behind this status is forgotten, and the temple of Artemis as Ephesus could very well be such a place. This temple holds the status of one of the seven wonders of the ancient world, it was destroyed in 401AD and was subsequently forgotten until it was rediscovered by British archaeologist and architect John Wood in 1869. Within that intervening time, even though the history of the site was forgotten the space came to be used by different religions as they came into popularity, thus continuing to respect the location as sacred even

1

though it was roughly 1km outside of the city and thus not as accessible as it could otherwise be.

Figure 1. A column reconstructed by Wood is all that now remains of the temple of Artemis (2011).

Artemis of Ephesus:

Just as the temple holds a special significance, so to does the nature of the divinity being worshipped herself. Artemis is a well known figure in Greek mythology, but the Artemis found here seems completely removed from the virginal huntress found in Greece. This goddess has a stronger connection to fertility, and so is more closely related to the Phrygian Cybele in her duties. She was not considered to be a sexual goddess like Aphrodite, but rather a faithful and nurturing mother as well as the legitimate wife of the city. She is understood to have been the patron deity of Ephesus prior to any influence by the Greeks, originally brought in by the Amazons whom she is known to protect. Once the Greeks arrived in Ephesus, they culturally appropriated this divinity by equating her with Artemis.

This form of Artemis is not simply a local reinterpretation of the goddess, but rather she can be found elsewhere in the Mediterranean wherever devotees have taken her. If this unusual reimagining of an Olympian goddess can retain followers even after they leave her immediate vicinity, then it must be due in some way to the strengths ascribed to her within Ephesus and the deep connections made between her and her worshippers.

Figure 2. A statue of Artemis of Ephesus, showing her Asian influences as well as the multitude of breasts tying her more closely to fertility (2011).

The διοπετής of Artemis:

All of these peculiarities are exemplified in Acts 19 with a confusing scene of St. Paul's missionary work in Ephesus. The message of this story, concerning a riot instigated by a silver smith defending the greatness of their goddess Artemis, does not seem to be clear, as it shows the power of this local pagan religion in the face of St. Paul's message.

This strength seems to be centered in the fact that Ephesus is the known to guard the glorious temple of Artemis, as well as the sacred stone sent to them by Zeus. If such a stone existed, it likely would have been placed behind the altar of the temple, which would have been made to house, protect, and celebrate this divine gift. A craftsman may have altered the stone in some way so as to make its connection to Artemis more clear, but his work on the stone would have been irrelevant in comparison to the meteorite's origins. St. Paul may claim that "a god made by hands is no god," but this statement would not apply to this cult statue because its religious significance is connected to the divine material rather than any mortal craftsmanship. The presence of this meteorite would have predated the temple, as the temple's placement would have been determined by its presence. The manipulation could have also predated

3

the temple, as the structure represents the Greek worship of this sacred stone.

Conclusion:

There may be no direct and conclusive evidence of a meteorite in ancient Ephesus, but the power and popularity of its cult of Artemis would suggest that its worshippers had a concrete relic to inspire such fervent devotion. The meteorite itself wouldn't have been the inspiration for such a regionally specific deity, but it would have been the item around which these new beliefs coalesced, making it central to the cult's success beyond its regional borders and appropriate time frame.

References:

[1] Brinks C. L. (2009) The Catholic Biblical Quarterly, 71, 776-794.
[2] Oakley K. P. (1971) Folklore, 82, 207-211.
[3] Parvis M. M. (1945) The Biblical Archaeologist, 61-73.
[4] Witetschek S. (2018) The First Urban Churches 3, 235-252.

THE IMPORTANCE OF PROPER REVERENCE FOR THE MAGNA MATER METEORITE IN ESTABLISHING THE IDENTITY OF AUGUSTAN ROME

Jilene Malbeuf[1] and Austin Mardon[2], [1]University of Alberta History and Classics department, Edmonton, Alberta, Canada, [2]Antarctic Institute of Canada.

Introduction:

In discussing the effect that Augustus' rule had on the overall appearance of Rome, Suetonius quotes as saying that "I found Rome built of bricks; I leave her clothed in marble" (Life of Augustus). With such import being placed on his various beautification projects as a tool of propaganda while he was in power, the choices he made in which buildings to revitalize show what kind of messages he was attempting to convey. One such choice was in remodeling the temple of Magna Mater which housed a sacred meteorite on Palatine Hill, the location where his own palace would rest. With such close proximity to his residence, any building on the Palatine would have an added significance for his overall image, so it is important to understand why he would go through the trouble of providing Magna Mater with a new shrine when he was already putting a great deal of effort into building his Temple of Apollo to commemorate his victory in the battle of Actium which ended decades of civil war and allowed him to rule uncontested for the remainder of his life.

The ruins of Palatine Hill as seen from the Colosseum in Rome (2006).

Magna Mater:

The deity Magna Mater, or Great Mother, was developed from and assimilated to the Phrygian mother goddess Cybele and the Minoan / Greek goddess Rhea who protected infant Zeus from his Titan father Cronus. For the Romans, she became associated with Troy, and as such was used to emphasize Rome's connection to the Trojan's. This emphasis was important as a way to connect Rome with the mythology of Greece while simultaneously separating themselves from their neighbors.

The meteorite that was the sacred relic of her cult was likely connected to her due to where it landed. Mount Ida had close links to both Troy and her cult center Pessinus (in modern day Turkey), so this is the most likely place where the meteorite fell, after which it would have been found and taken to Pessinus to be worshipped as the image of the goddess. The stone itself was used as the head of the cult statue, which also had a crown and rested on a thrown for her devotees to worship.

This particular meteorite has been lost since ca. 500AD, but its movements through Asia Minor to Rome were extensively discussed by the writers of its time, and its religious connections and uses are shown to be the typical reaction to meteorites in antiquity. While the stone itself as used as the image of a god, it is the stone's falling from the sky and any connected phenomena which inspires religious devotion in those who come to worship it. For this particular area, meteorites connected to some form of Cybele were housed and worshipped in Delphi and Ephesus as well, showing that this goddess's connection to Zeus makes her easily equated to a stone falling from the sky as his thunderbolts often do.

Magna Mater's Journey to Rome:

Although the Temple of Magna Mater grew in popularity under Augustus thanks to his revitalization of the site, the stone which was the reason behind this temple was actually brought to the city much earlier during the Republic, in 204BC during the Hannibalic Wars. At a point of uncertainty in the war, the Sibylline books were consulted for guidance, and the oracles told the Romans to bring the "Idaean mother" to Rome. They undertook this quest and brought the meteorite from Pessinus where is was being worshipped at the time into Rome, and as luck would have it Hannibal was forced into North Africa the following year.

As a sacred stone, this would not have been a simple matter of travelling into Asia Minor, picking up the stone, and going back home. Contemporary sources tend to disagree on the details of this process, but it is likely that the Romans used their friendship with Attalus the king of Pergamum to secure the stone as a gift, as a means of strengthening political bonds and displaying the strength of the Republic during tumultuous times. Rome had no ties with Pessinus at this time, which was the cult center for Magna Mater at this time, and continued to be so after the removal of the meteorite.

Magna Mater under Augustus:

With the Republic's victory over Hannibal as the Sibylline oracles foretold, Magna Mater's role as the protector of Rome was secured, but she remained a relatively minor deity with a quaint shrine until it burned down in 3AD. At that time, Augustus used the shrine's destruction as another opportunity to glorify the city with the added splendor of magnificent architecture. In rebuilding the shrine and keeping it so close to his personal residence, he incorporated the stories of Magna Mater into legends of his own lineage as well as Rome's foundation.

At this time, famous authors Ovid and Virgil were also writing their own stories which in one way or another promoted Augustan propaganda. Ovid told the story of Magna Mater being the protector of Troy and upon its destruction attempting to follow Aeneas to the new homeland. He also tells of how Claudia Quinta (an ancestor of Augustus's wife Livia) was instrumental in bringing Magna Mater's sacred throne out of Troy as far as Mount Ida. As Mount Ida was closely connected to Troy, Ovid insisted that this is where Rome procured her meteorite from, although this does not fit other more historically accurate sources.

Virgil's method of connecting Rome to Troy via Magna Mater is at once more mythical and more concrete. In Book 2 of The Iliad Aeneas is attempting to flee Troy amid its destruction with his son and father, but his father Anchises refuses to leave at first. It is only after he witnesses a shooting star fall on Mount Ida that he recognizes the sign from the gods saying that he must leave his home forever, and thus Aeneas begins his journey to what would eventually become Rome. After departing, their first stop is Mount Ida where Aeneas and his fellow survivors construct a fleet of ships out of pine trees from Magna Mater's sacred grove on the mountain. In these ways, Magna Mater is shown to be the protector of Troy, Aeneas, and Rome all at once.

It is clear that the scene with the shooting star would accurately reflect how the actual meteorite resting in the shrine would have fallen and become incorporated into the local religions. This correspondence would have added significance to the meteorite and shrine beyond what was typical of meteorite worship. In combining a material object to Rome's creation myth in this way, Augustus ensured that Magna Mater would remain a popular deity for centuries.

Conclusion:

Meteorites were a popular object to use for cult statues in antiquity due to their clear connection to Zeus and the heavens, but the meteorite of Magna Mater was particularly popular thanks to her additional political connections. Magna Mater was first established as the guardian and protector of Rome through a successful conclusion to the Hannibalic Wars, so even before her connection to Rome's Trojan roots was fully utilized her presence in the city was primarily a political move. Once Augustus came into power and sought to hold onto that power being what was traditionally permitted, her religious popularity grew rapidly alongside her importance to the Roman image as a New Troy. This Trojan connection allowed Rome to become a part of the Age of Heroes in Greek mythology, giving its citizens access to immortal glory. As first among the Roman citizens, Augustus was especially glorified by this connection with his ancestral connection to Aeneas, implying that the divine right to rule is in his blood even if Rome does not recognize a ruler. With access to a sacred meteorite that was allegedly present for the fall of Troy, this connection is solidified and Rome's immortal glory is secured.

References:

[1] Bell R. (2009) Papers of the British School at Rome, 77, 65-99.
[2] Burton P. J. (1996) Historia: Zeitschrift für Alte Geschichte, 45, 36-63.
[3] Farrington O. C. (1900) The Journal of American Folklore, 13, 199-208.

Comparing Newspaper Movie Reviews to Online Movie Reviews

Jessica Jutras

This rhetorical analysis essay, which is the second part of two papers, details the differences in structure, style, and audience between newspaper movie reviews and website movie reviews (whereas part one discusses three online reviews in-depth). The goal is to identify the differences between the two formats and how they affect the author's choices when writing. A newspaper review is compared to the three online reviews already analyzed. The analysis demonstrates that although the two formats share some similarities, newspaper reviews generally target different audiences and are structured differently.

Part I of this rhetorical analysis discusses three online movie reviews from 2012 to 2020 in which the writers follow the conventions of a typical form. Online reviews usually follow an etiquette wherein they discuss key factors such as a film's director, genre, or premiere. They also inform readers of the plot while keeping reviews spoiler-free; and lastly, online movie reviews often use popular (pop.) culture references liberally. Comparatively, the rhetorical form of newspaper reviews is different based on their dense tabloid size and targeted audience. Looking specifically at reviews from 2000 to 2010, when North America began noticing a steady decline in newspaper consumption (Lucena) and day-of-release film reviews (Halbfinger), Katherine Monk's 2007 review of American Gangster, published in the Edmonton Journal's "Friday Flicks" section, demonstrates the distinctions between newspaper and online movie reviews.

Monk's review, titled "Crime saga revisits American Dream," has some similarities to the style of web reviews, such as using a rating system, being categorized under reviews, mentioning the genre and director in her introduction paragraph, and detailing the plot spoiler-free. The differences in form, however, outweigh the similarities.

Newspaper reviews are short and compact at approximately 500 words per while fitting three to five reviews onto one tabloid-sized page, which often measures 260mm wide to 430mm long. Perfect for the casual reader skimming through pages on their way to work or in the waiting room. So, they concisely tell readers what the movies' themes are while critiquing them. Monk spends most of her review describing the main characters and plot points, then compares the motifs to Iraq, Vietnam, and American war conflicts and its relation to the "American Dream" ideal (par. 11). In contrast, online reviews have a full webpage to use, giving them more freedom in length (approximately 1000 to 1500 words) and discussion points. For example, Tasha Robinson has room at the beginning of her Joker review to detail the conversation surrounding the film's release (pars. 1, 2).

The targeted audiences of the two types of movie reviews differ. Newspapers primarily target a general audience. In contrast, web reviews aim to reach frequent readers who often want to gain more insight into the film before they see it (you may have been this reader: browsing the abundant selection of films to stream or choosing to buy movie tickets after arriving at the theatre, and relying on the Rotten Tomato movie reviews you read earlier that week). This audience differentiation affects both the length and depth of movie reviews and the expectations that writers may have of their readers. In a web review, writers may make pop culture references freely, assuming more from their audience, as exemplified in Peter Travers' review of Tenet.

However, in a newspaper review with a broad audience, writers cannot assume as much from their readers. They want to attract a variety of people to the paper, so they typically only expect readers to have a rudimentary understanding of the genre or recent movie releases; for instance, Monk expects readers to understand her reference to the common "oil-slicked . . . cops 'n' mobsters" convention of 1970-styled crime films (par. 3). Even while referencing other movies, it is most often brief and relative to a critique, director, or actor's experience. Namely, Monk claims that American Gangster is director Ridley Scott's best production since his movie Gladiator (par. 9).

In sum, despite sharing the handful of similarities listed, Monk's review demonstrates that newspaper movie reviews follow a rhetorical format that differs substantially from online reviews. Their compact layout and broader targeted audience affect the differences in form the most, restricting writers to only necessary and immediately relevant information for readers.

***

Jessica Jutras is an undergraduate student at MacEwan University's School of Business, studying Library Information Technologies. She is the author of three books, including Music and Mental Health: Let's Talk About Emo.

Works Cited

Halbfinger, David. "Day-of-release film reviews a thing of the past." Edmonton Journal, 6 July 2007, https://search-proquest-com.ezproxy.macewan.ca/hnpedmontonjournal/docview/2403958453/4821F2A3A27846 ACPQ/1?accountid=12212

Lucena, André. The Print Newspaper in the Information Age: An Analysis of Trends and Perspectives. 2010. University of Alberta, Masters dissertation. ERA, https://doi.org/10.7939/R3CF14

Monk, Katherine. "Crime saga revisits American Dream." Edmonton Journal, 2 November 2007, https://search-proquest-com.ezproxy.macewan.ca/hnpedmontonjournal/docview/2403894149/pagelevelImagePD F/9398ABAFC2764805PQ/2?accountid=12212

Robinson, Tasha. "Love it or hate it, the Joker movie presents a tempting fantasy." The Verge, 4 October 2019, https://www.theverge.com/2019/10/4/20899422/joker-movie-review-toddphillips-joaquin-phoenix-incel-violence-dc-comics-batman.

Travers, Peter. "'Tenet' Review: Christopher Nolan's Knockout Arrives Right on Time." *RollingStone*, 28 August 2020, https://www.rollingstone.com/movies/moviereviews/tenet-movie-review-christopher-nolan-1047641/.

Alcohol Use Disorder

Chitrini Tandon, Muzammil Syed, Dr. Austin Mardon

Alcohol Use Disorder (AUD) is a psychopathological disorder that impairs the frontal cortical function of the brain and can affect individuals for prolonged periods. In the USA, AUD affects approximately 15 million people each year (2018). It can be caused due to a variety of reasons including but not limited to problems controlling your drinking, continuing to drink even after negative symptoms arise, and binge drinking (2018). Relapse is unfortunately common and is caused by exposure to contextual cues that trigger memories of drinking and increase cravings (2018). Furthermore, studies have shown that it is the seventh leading risk factor in death and disability-adjusted life-years worldwide (Moggi et al., 2020). According to the Centre for Disease Control (CDC), AUD has an economic burden of over 250 billion dollars in the USA per year (Coker et al., 2020). Unfortunately, despite the devastating impact of AUD on the patient and healthcare level, AUD has been studied but there are still many gaps in the research due to alcohol's complex molecular actions and the two types of alcoholism, each with their symptoms. Thus, a major knowledge gap exists regarding how AUD affects our brain and bodies.

Many signs indicate someone has AUD. Some of these include drinking alcohol in unsafe situations (e.g. drinking before driving,) failing to meet daily life obligations such as going to work and decreased participation in social/work/personal activities, among others (2018). There are also a variety of symptoms patients with AUD may experience, such as difficulty in decision making and impulsivity (due to discounting of delayed rewards), high levels of psychoticism such as aggressiveness, high levels of neuroticism and low levels of extraversion (Moggi et al., 2020). Depending on the number and intensity of these symptoms AUD can be classified as mild, moderate or severe. Other cognitive deficits include visuospatial functions, attention and visual-perceptual motor processing, learning, memory, simple motor controls including balance distribution and executive functions such as abstraction, problem-solving and cognitive flexibility (Moggi et al., 2020).

There are currently three Food and Drug Administration (FDA) approved drugs on the market for AUD, namely disulfiram, acamprosate, and naltrexone (Coker et al., 2020). However, each of these drugs has its disadvantages and drawbacks. Disulfiram is an irreversible inhibitor of alcohol dehydrogenase; it primarily works by inducing unpleasant physiological effects when combined with alcohol, this drug is only effective in compliant patients (Coker et al., 2020). Acamprosate is a positive allosteric modulator, whereas naltrexone is a non-specific opioid receptor antagonist (Coker et al., 2020). While they work differently, both are effective in increasing abstinence and reducing alcohol intake. However, they still have their side effects, which may decrease patient compliance and limit treatment efficacy (Coker et al., 2020). Thus, despite having received FDA approval the side effects associated with these three drugs render them seldomly prescribed.

However, not all is doom and gloom. Scientists are actively researching AUD with many recent discoveries about better diagnostic methods of the disease, as well as enhancing our current understanding of the mechanisms involved with how alcohol affects our brain. For instance, one recent study suggested that catechol-O-methyltransferase (COMT) inhibitors such as tolcapone may be effective in therapeutics for AUD (Moggi et al., 2020). Similarly, other studies have posited two different models that may potentially explain the development of uncontrolled alcohol consumption. The first being Type I alcoholism (passive dependent or anxious personality) - these are people who exhibit low levels of novelty seeking and high levels of reward dependence and harm avoidance (Moggi et al., 2020). The second being Type II alcoholism (antisocial personality) - these are people that exhibit high levels of novelty seeking and low levels of harm avoidance and reward dependence (Moggi et al., 2020). Another study was able to find a link between context-specific regulation of transcription in cue encoding neurons and the lasting effect of ethanol on transcript usage during memory formation (Petruccelli et al., 2019).

Despite these recent discoveries, there are also many unknowns surrounding AUD. For instance, the relationship between the DSM-5 maladaptive personality domains on the comorbidity between AUD and different personality disorders (PDs) is yet to be fully determined (Moggi et al., 2020).

In conclusion, AUD is a complex psychopathological disorder, and much is yet to be learnt about the specific mechanism and pathways of this disorder. While there are current treatments available, further research needs to be conducted to find more effective therapeutic options for patients diagnosed with AUD.

Works Cited:

Coker, A. R., Weinstein, D. N., Vega, T. A., Miller, C. S., Kayser, A. S., & Mitchell, J. M. (2020). The catechol-O-methyltransferase inhibitor tolcapone modulates alcohol consumption and impulsive choice in alcohol use disorder. Psychopharmacology, 237(10), 3139–3148. https://doi.org/10.1007/s00213-020-05599-5

Moggi, F., Ossola, N., Graser, Y., & Soravia, L. M. (2020). Trail Making Test: Normative Data for Patients with Severe Alcohol Use Disorder. Substance Use & Misuse, 55(11), 1790–1799. https://doi.org/10.1080/10826084.2020.1765806

N/A. (2018, July 11). Alcohol use disorder. Mayo Clinic. https://www.mayoclinic.org/diseases-conditions/alcohol-use-disorder/symptomscauses/syc-20369243.

Petruccelli, E., Ledru, N., & Kaun, K. R. (2019). Alcohol causes lasting differential transcription in Drosophila mushroom body neurons. Genetics, 215(1), 103–116. https://doi.org/10.1101/752477

***

Chitrini Tandon is an undergraduate Life Sciences student (McMaster University) is an author, and disability advocate. She wrote this article in association with the Antarctic Institute of Canada Team.

Austin Mardon, PhD, CM, FRSC is a publisher, writer, advocate for mental health, and speaker based in Edmonton, Alberta, and the director of the Antarctic Institute of Canada. He is an assistant adjunct professor at the John Dossetor Health Ethics Centre at the University of Alberta.

Muzammil Syed is a Masters student at the University of Toronto.

Gender and Migration

Ananda Majumdar

The University of Alberta (Bachelor of Education after Degree Elementary, Faculty of Education,

Community Service-Learning Certificate and Certificate in International Learning, CIL) *

Harvard Graduate School of Education (Professional Education as a Child Development

Educator, Certificate in Early Education Leadership (CEEL-Series 2), online) *

Prospective Summer School in History and Archeology (2021-22)- "On the footsteps of Jesus: Jerusalem to Magdala, European University of Rome, Italy

Certificate in Migration Studies, GRFDT, New Delhi, India (In Progress, Online)

Grant MacEwan University (Diploma in HR Management)

Jadavpur University (Master of Arts in International Relations)

Sikkim Manipal University (Master of Business Administration in HR and Marketing Management)

MBB College, Tripura University (Bachelor of Arts in Political Science)

Antarctic Institute of Canada (Researcher and Writer), Servicing Community Internship

Program (SCiP) Funded by the Government of Alberta

Member of Student Panel, Cambridge

University Press, Member of the Associa-

tion of Political Theory (ATP) University of

Massachusetts
Student Member of ESA (European Studies Association), Columbia University,
U.S.

General Coordinator, Let's Talk Science, University of Alberta

Early Childhood Educator, Brander Garden After School Parents Association

Education Support Staff, Brander Garden School

Reviewer, Journal of Education, Society and Behavioural Science, ISSN: 2456-
981X

Cell# 1-780-660-7686,

anandamajumdar2004@yahoo.co.uk, anandamajumdar2@gmail.com, anan-
da@ualberta.ca

Abstract:

Gender is an important term in migration. This term has been used for the relocation of women outside and inside of their country for their betterment of life, rights. The objective of the unit article is to know the gender in migration especially about women migration outside and inside of their country. It is some knowledge about how women spend their life aboard. How do they face various problems and become vulnerable? The outcome of the unit is a complete research on gender issues in migration and discover protection policies to maintain a standard of migrants regardless of status. The unit has been written through various websites academic articles and articles from international organizations such as the United Nations newsletter etc. It is therefore a documentary analysis based on a qualitative approach. Gender in-migration has advantages and disadvantages. It has expanded the mobility of human beings by adding women relocation and thus provides women with some power to become independent and challenge traditional society. It has provided an opportunity for social and economic and skilled development through exchanges, migrated,

17

emigrated, and immigrated. This is how flexibility and collaboration have been built in a neoliberal world. Talented migration creates brain drain in original countries, and a gap to invest in demographic structures such as education, health care sectors. It creates a gap of intellectuality that affects a country's development. The feature question is how does feminization of migration buildup and why?

8.1 Introduction:

For better and opportunities people are moving around the world, around the country. This is how they are immigrating inside their country, emigrating permanently to other countries, especially developed countries, and migrating temporarily as a worker to fulfill the work labour force in the middle east countries. They are immigrating or emigrating or migrating on their wish, hope and for family development, self-development. They are also moving due to forceful ways which are forced migration. This is because of natural disasters, political, social, and economic situations, and because minority disparity. It is a fact that migration experience can be shaped by sexual identity and orientation, gender. Gender sways the reasons of migration such as various narratives of migration; how and where to migrate? Why migrate? What are networks for migration? Who migrates? Gender and community differences also shape the risk, vulnerabilities of migration. Features like the role of expectation, power and relationship and vitalities are depended on being a man or woman, boy or girl and different sexual identities such as lesbian, straight, gay etc. they affect the migration process and be affected in various ways or new ways by the policies of migration. Therefore, it is important to aware that how do migration and gender impact each other and how are they interrelated and on the other side respond appropriately. When migration defects by the naturalization of gender, an international organization for migration starts working to set up a balance or equality between men and women, boys, and girls etc. IOM then works by advocating equal rights under the law in employment and mobility, fighting against gender discrimination, responses that how gender impacts social works, economic growth, risk and vulnerabilities and migration responses on gender rules and relations. Gender now an important feature of international migration. Research in the 1970s and 1980s began to add women in the migration diaspora. Various features, the environmental situation depends. The demand for labour in receiving countries. The migration of women domestic workers is demandable in North America, Middle East, and Europe. Feminist theory also provides importance on gender-based migration. The theory believes in gender as a medium of behaviour, power, and relationship. Feminism says that gender is a social structure, a social construction. Feminism in the '80s and '90s has focused on gender rather than men vs women. Gender is a much-weighted word than men vs women. It motivates migration and related processes, such as the adaptation of a new country, the continuation of livelihood in a new country and overall outcomes. Feminist view of gender as a social construction has raised various questions. The question is based on the power of patriarchy wherein the developing countries men have several sources to access to control women; in this situation how do women able to migrate? How do interpersonal relations between men and women impact women's migration especially when the family matter arises? Women are responsible for family care, in this situation is there any impact when they migrate to other countries for labour work? After reading these themes gender issues will be easier to realize.

1. Definition of gender and migration.
2. The collaboration of gender and migration and gender in migration.
3. Causes of Women relocation.
4. Migration policies and initiatives are taken by an international organization.
5. A term called Feminization of migration.
6. Pros and cons of migration.

8.2 Meaning and concept of Gender and Migration:

Gender has been defined by the International Organization Migration (IOM). The organization says that gender is values, attitude, power, influence, relationship; through which society can assign people based on their sexual orientation. Therefore, gender is a social contract that has roles and relations among the citizens. It is not only referring to men and women but other groups that as well can relate, among others. Gender has variation in every culture, between and within culture; and is deeply entrenched with culture. United Nations says gender is not only women, but it also refers to men, and other gender crowed. In general discussion, gender suggests on women because of socio-economic gender inequality, intersectionality. But gender as a group of other sections can bring equality. According to IOM gender does not only refer to sex they existed during their birth. Gender is a personal sense of the body that includes speech, dress, and behaviour. According to WHO gender is a narrative that describes the features of men and women, is also socially constructed. WHO (World Health Organization) says; sex refers to a biological feature that reflects by boys and girls and later they turn into men and women? This is a lesson behaviour that makes a stance as gender identity. Gender identifies various sectors such as inequalities, intersectionality etc. and discusses health by focusing on discrimination over women and how does it affect women's lives? Gender has also discussed the solutions. It says women face to attain health and its speech can solve the entire issues of inequalities. The concepts of gender are various, such as gender equality, equity, inequality etc.

Migration on the other side has been defined by the UN. It says migration is people's movement. From the earliest time humanity moved from one place to another for shelter, food, other economic opportunities, join family etc. Migrant is an important term in migration, has been described by the IOM (UN Migration Agency), says that migrant is any person who individually or as a family moved outside of the country as an immigrant; or moved within states of their country as an immigrant, or migrant as temporary labour(migrant). In the present world, people are living other countries than their original destination. In 2019 estimated 272 million people were migrant status globally. It was estimated 10 million more than in 2010. This is how the globalization of migration started worldwide. It was not only workforce migration temporarily, but it was also a migration of brain drain through talented professionals, students in the developed countries (general migration). Through the advent of globalization, the West opens their border for the demand of its workforce and bright professionals and students. This is how people from developing countries especially from South Asian regions migrated to the workforce temporarily for their better economic life and family.

8.3 Importance of Gender in Migration:

Migration turned into a universal reality that is occurring worldwide with settlement, mechanization, and development. Feminization in the workforce has been started after the world war 2nd as its impact.

20

The thoughts of migration have been expanded due to the beginning of the global economy. Migration is misinterpreted as a male prodigy mostly and women were dependent on men. Women were diminished in such away. The figure of 2002 showed the numeration of women migrants as male ones. However, the data of women migrants are not that noticeably clear as it was always a hidden feature. Donna Gabaccia has told at a seminar on families on the move that international policies have failed to detect or recognize female migrant labour. There are only 22 countries that have been concerned with the International Labour Organization (ILO) about female migration and the rights of female domestic labour against exploitation. It has been observed that women in other countries as a temporary migrants can not work independently and they also report on abusive attitudes by the employers and others. It is therefore an observation that women's migration has not been yet tumbled under the legal system. Migration in South Asia is ruled by basic factors such as family, agricultural style of manufacturers, and weddings. Rural-urban migration is an economic motive. It is for a better livelihood and comes out of poverty. Traditionally, especially in the South Asian region men migrate to the city from the village for a better income and his women wait for him for the return and utilize the funds carefully. In the Indian scenario wedding is a social contract where women work as labour in their marital home. Her domestic, agricultural, and miscellaneous labour is valuable to her husband and his earnings to her. The migration is therefore a decision by the couple. Economic migration is the most effective way for internal or external people's mobility. Forced migration such as war leave, conflict-ridden areas where there is no choice but to move etc. is another migration where gendered issues are frequently skipped. The Rohingya crisis is a great example in the context of female vulnerability. Those female refugees have been raped by the Burmese army. Rohingya women have been raped and became pregnant. They are not aware of abortion and thus they abandon their children, mostly female babies. The measurement of sex trade falls under forced migration. Smugglers abuse refugees and migration seekers by saying they will help them for the moving to the developed world. Smugglers ask for large sums of money but the inability of paying money results in sexual assaults and prostitution as reimbursement. Equal pay is another movement for the fairness of women's wages compared to men. There is a piece of common information that there is a substantial wage gap between migrants and the domiciles of the country. Most blue colour migrant workers are employed through a formal contract. There is no trade union, resulting in a more stressful life creates by the bad employers.

8.4 Gender Mainstream Migration Policies:

Understanding migration in the context of a gender perspective is a way that can provide guarantee and protects the rights of all gender identities. Integrating a gender mainstreaming approach to migration policies is important. These migration policies are connected with the 2030 agenda and with sustainable development goals such as protect labour rights with promoting safe and secure working place for all workers including temporary migrants' workers, specifically women migrants who are in risky employment; eliminate all forms of violence against women and girls in the public and private sectors they work etc. improvement has been observed in North America, Caribbean and Central American countries to thinking about gender mainstreaming cooperation that can be linked with migration policies. Regional Conference of Migration (CRM) is a regional consultation forum where gender issues discuss in migration issues. They prioritize gender on migration issues. It is a forum where women migration is a primary topic on their discussion table since 2017. The forum has an assistance program for women's development and as a helping hand on women's migration. They have focused on an issue that how does gender focus on inequalities in migratory courses. The forum explains those policies that adopt a range of perspectives such as intergenerational, intersectional approaches, human rights etc. Forum of President of Legislative Powers in Central America and the Caribbean Basin (FOPREL) is another forum that focuses on migratory issues, reports, various assistance programs in the region. They make a policy that is linked with the legal framework on the matter of migration along with the human rights approach. in this policy, framework gender has been prioritized such as a special reference to migrant women, youth, teenagers, and girls. The forum was published in 2019 to provide some advice on promotional safety, regular migration for susceptible people of Central America, the Caribbean and Mexico. There are special services that have been recommended for the well-being of the migrant populace such as medical care, legal, mental health service assistance etc. these are few things that ensure to protect the victims of violence. Inclusive migration management is another way that can make policies for the equalities of gender in a migratory environment. For a better and smooth migration record, implementation of standards, programs, strategies is necessary for indigenous women, migrants with disabilities. Another way to add gender in migration is research and data collection. It ensures correctness and avoids mistakes. It also ensures positive and focused research on migration development and positive impacts on it. Communication and campaign are other ways for the awareness of gender in migratory development. Digital media(policy) is a positive campaign as a medium to aware women on procedures, services, support programs, health risk during migrating, the suggestion to adopt a new country, workers recommendation etc. The forum also provides a mobile application that can help women to submit their migratory questions and to be directed to the right institution. Migrant women face multiple discrimination based on race, ethnicity, religion, sexual orientation, disabilities and so on. Migration policies should be recognizing women's agency and provide power to women for the reducing of inequalities. Policies should work on varieties of platforms that can help to access women in multiple ways. The state is itself an institution and liable to promote an environment where the state helps to promote, protect, and guarantees the rights of all diverse migrants in the region.

8.5 Women in Migration System:

Gender dimension in migration is its characteristics. Migrant women are a theme through which the realities of women's exploitation by migration come out. There are various international laws and treaties on a legal framework for people on the move, but those legal frameworks do not fit women's issues. People on the move have been categorized as internally displaced persons, internal economic migrants, international migrants, refugees, or asylum seekers. The state must fulfill the human rights and treaty obligation. International human rights law confirms rights for all regardless of any status. It should address the multiple and complex reasons for women's migration, uphold the human rights of all women in migration and fulfill the demand of human rights to every woman in migration regardless of any specific women category. International human rights also ensure a policy that implements and develop a parallel global compact on refugees. The human rights law promotes and develops a policy to protect the rights of women at all stages of migration. The UN women is a great example here. Women in migration are not creating any violence but they are advocating for their rights. Immigration policies currently have no clear dimension about women; it has inequalities and several vulnerabilities that can not protect women in other countries where they are going to migrate for economic reasons or by force. Migration policy should be implemented based on human rights, gender-responsive and without delay. This policy accepts women as an agency of migration. The global compact should implement a policy that can create an independent status for women in migration, women access to individual documentation etc. Women are always victims of various abuse and oppression, especially women in other countries as a migrant. This is how intersectionality creates when women have been faced multiple identities at the same time and face multiple coercion. Racial discrimination and Xenophobia are dangerous for migrants, and immigrants in other countries. Policy changes at the national level can create some protection. The Global Compact should encourage the end of all oppression. Women migration is a push factor. Push factor discovers through economic reason. For a better economic life or family lives women migrate. Women including Lesbian, Transgender are experiencing gender-based violence, organized crime, institutional oppression etc. The Global Compact in this context should endorse rights-based development that can flourish women's human rights, women's role in decision making, women's bodily integrity, social and economic protection, and decent work for women. Outlawing migration, firewalls are common which is implemented by global leaders and policymakers. Every state has its power to control the border. Through nobody can completely deny global human rights policies. The criminalization of migration does not count undocumented immigrants. Un-documentation is not a crime. Women face various types of bans, violence. Migration policy should create in a way that can secure women as much as possible and secure every migrant. The Global Compact in this context ensures an independent residency status for women which is secured for them. The policy should encourage a permanent resident program from the status of migrants. The policy should also make a firewall between justice, public services, and migration application. International orders can not be an open or exception for anyone. Migrants cross border always hope a safer border check. In the context of Women in transit, it may be a danger at the international border where exclusive attention is required. During cross border check various things happen due to irregular activities. The state should confirm migrant's safety and address their necessity. The state should not facilitate borders through militarization the prevent

24

migration. The global compact in this context should deliver access to justice and due process of law for women in all areas and stages of relocation. The policy should ensure that border governance measures conform of dignity and humanity to all migrants during entering and departure. The policy should ensure that asylum seeker and refugees entering other countries through land, air or water as a border cross is safe. The policy should be aware of immigration detention and commit to ending all forms of immigration detention and maintains a global human rights law as technical protection. States are always careful about smuggling(trafficking) in person and they restrict persons includes women due to this logic. This is a dangerous issue in the context of movement globally. However, this harm migrants' rights, increase criminalization and imprisonment. Regular channels of migration would reduce any smuggling networking. The Global Compact in this context should define trafficking in persons and call on the state to fulfill their obligation. The policy should ensure that anti-trafficking measures do not affect the human rights of all humans regardless of status. It is especially important to understand that migrants are not criminal who uses the services of smugglers; therefore, immigrants should not be criminalized. In the context of labour migration, every human being has the right to work regardless of immigration status. Unfortunately, many countries do not comply with this ethics of international human rights law or the charter. Women migration is more irregular, not systemized, informal, which places women in various difficulties such as women are the most victims of lower-wage, lack of opportunities etc. and they are facing exploitation. In this context, the Global Compact should abolish job separation by gender. The policy should include workers' voice in co-operation. The policy campaigns for the role of ILO and the development of migration and development based on ILO's procedures. Therefore, global cooperation is welcome for a safe attachment between gender and migration. It is an urge that focus should be on a global community that cares about migration governance. Deportation, imprisonment are few things in a cross border that needs to be handled. It needs to be remembered that the Global Compact is not a binding instrument but to follow for a better world where gender equality exists. In this context, the Global Compact should arrange meaningful participation of civil society such as migrant women organization. The policy should ensure dedicated funding for the migrant-led and migrant women organizations.

8.6 Feminization of Migration:

Feminization of migration is a multidimensional prodigy. An estimated 48 percent of women migrated for work and a better life in 2015 globally. The growing demand for women workers raises for women migration as a labour force. Fields are many such as care industry, domestic and manufacturing were demanding of women workforce are high. Women of the global north are not willing to work in those areas while women from the global south have the opportunity for higher wages than their original countries. Women are changing in the developing world. They are seeking freedom socially, economically. It is now rare that women move for family reunification but for jobs as they would like to be economically solvent. Women progressed estimated $300.6 billion in 2016, which is equivalent to half of the global remittances. It is a great performance by them like men. The year 1980 was significant for the changing of the world economy, since then feminization of migration was a popular term for the encouragement of women for their livelihood, to provide them freedom. Global South started privatization and they announced major cuts in social spending. They started to follow neoliberalism, which caused various problems in social demography in their countries. Higher poverty levels, social inequalities, unemployment rates; are increasing. It was difficult for men to work regularly. Women come to work for the assistance of her family, and they faced various gender superiorities and disparities. Inequalities of wages in different countries is another reason for female migration. In Mexico per day minimum wage is $4.00 while in New York state it is hourly $10. It has been said by the scholars like Castle and Miller that migration is the only solution to a public problem. It is a solution for individuals and families, it is also a solution for the government due to remittance. The care industry in the global north has increased the reason for migration. Childcare, elderly care are major areas in developing countries where women can be employed. These are the areas where migrant women can get the job. According to Cynthia Enloe manufacturing industries have been feminized such as garments, food processing, textiles industries etc. women migrants from the Philippines are working at Tim Hortons, McDonald's in Canada enormously which is a feminization of food industries in North America. Employers are hiring female migrants than male whether domestic or foreign workers due to their hard work, reliability etc. This is how the demand for migrant women is increasing. In its origin countries, women are controlled by the men dominated society, they are sacrificing for their household for nothing. This is another reason that women are migrating to developed countries for better opportunities, freedom and the happiness of her life. Countries like the Philippines, Sri Lanka have made especial agency for the promotion of female migration. Sri Lanka government has a pre-departure program to educate women on how to manage prepare food and arrange a table. The Philippines government has signed with various countries on the health care sector for the supply of women as a health care provider. Migrant women are the most vulnerable community in their destination countries due to their documentation status, which is mostly not clear or legal, their inefficiency in language, no idea about employment protection. It is called human trafficking. Women are working long hours for low wages, and works are humiliating. They are abused mentally, physically. The Nanny chain is one of the high work sectors for migrant women. They are providing surplus love to other children as a Nanny. They leave their children back home and migrates to their destination countries for hope. For the protection of migrant women treaties between countries should be adding protection of rights of all migrants' workers and their members of families according to international convention, which has been adopted in 1990 by the United Nations General Assembly for the protection of migrant's work as a written document. The Migrant Worker Convention guarantees the protection of migrants regardless of their status whether documented workers and undocumented workers. The protection is by the State in the areas of fair hearing, medical care, emergency medical care, fair hearing, consular services etc.

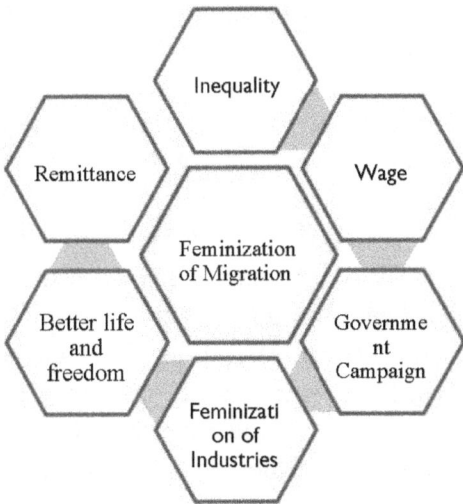

This is a brain drain for their origin countries. The government should create a job sector, better wage packages for the reduction of the feminization of migration. The government of original countries should invest in their health care sector. The government of original countries should invest in local economy such as education, infrastructure, public services for more attraction instead of campaigning migration opportunity. This is how brain drain can be controlled in the global south countries. Women who are yet choosing to migrate should create laws such as the Domestic Workers Bill of Rights of New York State, which provides minimum wage, overtime pay, a day of rest every week etc. to workers regardless of their status whether documented or undocumented. Feminization of migration is a campaign by the developing countries for the encouragement of female migration to the developed countries. It is a push for migration and putting women at risk of being abused, exploited by the transition. Therefore, implementation of policies is required to protect the rights of migrant women, along with their right to have a family, decent working conditions, physical integrity, fair wages, education and equal access to law and justice.

8.7 Let Us Sum Up:

India, Bangladesh, Sri Lanka, Philippines are countries where migration in the workforce has been given priorities for the remittance of the government of those countries. Gender has been an important narrative in migration because of its both characters (men and women) emigration, immigration, and migration inside and outside countries. Gender has been evaluated through values, respect of individualities whether girls or boys who become men and women. Gender is not therefore only female. Because of individual liberty, life enjoyment, better opportunities males and females both have been moved from one place to another place whether inside of the country or outside. It is a weighted word than men and women. Gender has been added in migration for the betterment of the female section, especially from developing countries. Women need freedom and individual rights; therefore, gender has been added for the protection of women through their mobility instead of the family server. International Organization for Migration (IOM), ILO (International Labour Organization) are an organization that has promoted the narrative of gender in migration. Humans move globally is for the flexibility of lives, this is how migratory decisions always have been taken by the couple, individual form a situation where rights have been exploited, lives have been risked and livelihood has been tolerated. Migration has been done by various ways such as immigration within the country for the labour force, emigration from one to another country by force, which is forced migration such as for political, social, and economic turmoil, natural disaster etc. migration temporarily from one to another country such as South Asian labour force to Middle East countries and other developed countries for the demand of health care, service, and elder care industries. Globalization and neoliberalism were ideologies from the developed countries that opened their border for the global south. This is how they hired talented people as a student, skilled workers and keep them permanently for the economic and social contribution of their countries. Canada and the United States are economically and socially contributed by immigrant people. They are a developed nation due to the immigrant population and their heavy contribution in cultural, economic, political, demographic characteristics. The settlement process has occurred through undocumented migrants, refugees and asylum seekers who have been fled to escape their lives from the political and social brutalities from their original countries such as Salman Rushdie, Dalai Lama; great people of their original country had given shelter for survival. Migration is a part of the global governance through its economic demands. Health Care, Elder Care, Food Processing are those manufacturing industries in the Global North where migrants are highly demandable. Employers are hiring women migrants and other migrants than their domestic citizens because of hard work, trust, and faith. Migration policies have been implemented by a global organization such as IOM, UN general assembly meeting on migration and implementation of policies by member countries for the protection of people mobilization, their rights etc. migrants' people are always minority section of every country. Racism, intersectionality has been raised, the movement has been raised through nationalism and for the deportation of illegal migrants. It is also a part of gender and migration. Policymakers were supporting the anti-immigration movement for political gain in their country. The pandemic world has shown the reality of the migrant's people everywhere. They are treating badly; they are deporting due to the fear of COVID-19. Their accommodation is in a worse situation, but they are still living due to no other ways. For the protection of those minorities, deprived population, especially women migrants' section who are most abused, international authorities have set up policies for their protection such as following the charter of human rights by every States and providing minimum accommodation, health care facilities, fair hearing, and consular services to protect from legal error. Migration policies have provided women to become powerful in their society and other countries for their knowledge and protection. It is a defence against any kinds of disparities from their household and the male-dominated society. This is a power to protect them in working places they are working in abroad. Women(gender) in migration reflects their rights, not violence. It is also a reflection that women are no more surrounding their house, sacrificing their life for the household services, for their husband and another family member. Criminalization of migration, human trafficking, are terms that create difficulties for the migration of women and others in the name of law and order of every country. Harassment is continuing in the check post during entering countries of destination. Therefore, global human rights ensure few policies that implement and develop a parallel global compact on migrants, undocumented migrants, and refugees. Human rights law promotes policy that can protect women regardless of their status in all stages of migration. Since the 1980's women have been a part of migration for their development, integrity and today estimated 48 percent of total migrants are women from around the world. Women from South Asian countries have been relocated to the Middle East countries, Malaysia for work in the garments industry, services etc. they temporarily migrated and returned as well. They face tremendous amusement, but they choose to migrate to come out of South Asian societal domestic women abasement. The government has launched various programs for the campaign of migration and to earn. It is profitable commerce for the government through remittance. The wave of women relocation domestically and globally has been recognized as the feminization of migration. In 2000 1 in 35 people was an international migrant where half were women. It is a way that led to a greater degree of economic and social autonomy for women. Migration is a way through which women can challenge traditional and restrictive gender roles in developing countries, especially in the Indian Subcontinent. It is a way where men and women both can earn higher wages and gain high training that makes them skilled, and few of them even send back their countries as a remittance. From the perspective of developing countries both by men and women migration increases economic output. Economic output has been reflected through living standards. Immigration leads to higher economic growth with the rise of tax revenues. Migration through men and women (from a gender perspective) makes them both entrepreneurs. They both arrive in their original countries with wealth which makes a potentiality to become businesspersons. Not only their original countries, both men and women immigrants get jobs abroad (UK, U.S.A) and they expend money in the market as well for their livelihood. It creates the demand for products and increases the market of those countries. Therefore, immigrants are not a lump of labour fallacy but assets of those countries. In the United Kingdom and the United States immigrants are highly skilled and educated and are working. It has increased the skilled labour force as their working asset. Immigration in the global north is good for the young generation. In the west old generation is increasing, therefore young immigration is a hope for those countries for the operation of production and young livelihood generation for the future asset. Young skilled migration is a hope to reduce skills shortages. Immigration whether men or women make a multicultural society in the global north countries. However, there are various disadvantages of immigration as well. Brain -a drain is one of the most disadvantages of original countries. They are losing talents. They are not investing in their domestic market for more demand and rises of employment, but they are campaigning and encouraging immigration to the global north. This is unexpected. Native-born people of host countries are losing jobs because of migrants' people which is called structural unemployment. Migration refers to the rise of the population through which pressure increases in the public services. However, for a running global economy, skilled migration through men and women is a must. Without skilled workers from the global south, the developed products of the global north countries could be in trouble.

8.8 Key Words:

A. **Feminization of Migration:** Feminization of migration is a multidimensional prodigy. An estimated 48 percent of women migrated for work and a better life in 2015 globally. The growing demand for women workers raises for women migration as a labour force.

B. **Migration Policies:** migration policies relate to the 2030 agenda and with sustainable development goals such as protect labour rights with promoting safe and secure working place for all workers including temporary migrants' workers, specifically women migrants who are in risky employment; eliminate all forms of violence against women and girls in the public and private sectors they work.

C. **Gender and Migration:** IOM says that gender is values, attitude, power, influence, relationship; through which society can assign people based on their sexual orientation. Therefore, gender is a social contract that has roles and relations among the citizens. Migration on the other side has been defined by the UN. It says migration is people's movement.

D. **Gender in Migration:** Feminization in the workforce has been started after the world war 2nd as its impact.

8.9. Progress and Possible Answers:

How does feminization of migration buildup and why? Feminization of migration has been built up since 1980 in developing countries. The percentage of the relocation women was quite high among migration. Industries like food processing, garments, elder care were highly dependable in the global north countries and the employers were willing to hire migrants' women due to their trustworthiness and hard work. It was a better opportunity for economic prosperity. The feminism of migration is a symbol of women's solidarity that has given power for the societal protection and have that power. It is a freedom from exploitation of women in developing countries.

8.9.1. References:

1. Grieco, E., & Boyd, M. (2003). Women and Migration: Incorporating Gender into International Migration Theory. migrationpolicy.org. Retrieved from https://www.migrationpolicy.org/article/women-and-migration-incorporating-genderinternational-migration-theory/.

2. Gender and migration. International Organization for Migration. Retrieved from https://www.iom.int/gender-and-migration.

3. Gender and migration. Migration data portal. Retrieved from https://migrationdataportal.org/themes/gender-and-migration.

4. Gender: definitions. Euro.who.int. (2002). Retrieved from https://www.euro.who.int/en/health-topics/health-determinants/gender/gender-definitions.

5. Migration. Un.org. Retrieved from https://www.un.org/en/sections/ issuesdepth/migration/index.html.

6. Godbole, D. (2018). Why Gender Is Important When Discussing Migration | Feminism in India. Retrieved from https://feminisminindia. com/2018/10/04/gender-acknowledgedmigration/.

7. Astles, J. How to gender mainstream migration policies. Regional Office for Central America, North America and the Caribbean. Retrieved from https://rosanjose.iom.int/site/en/blog/how-gender-mainstream-migration-policies.

8. Women in Migration Network. Refugeesmigrants.un.org. (2017). Retrieved from https://refugeesmigrants.un.org/sites/default/files/ women_in_migration_ts3.pdf.

9. Maymon, P. The Feminization of Migration: Why are Women Moving More? Cornellpolicyreview.com. Retrieved from http://www. cornellpolicyreview.com/thefeminization-of-migration-why-are-women-moving-more/?pdf=3479.

10. Jolly, S., & Reeves, H. (2005). Gender and Migration | BRIDGE. Bridge.ids.ac.UK. Retrieved from https://www. bridge.ids.ac.uk/bridge-publications/cutting-edge-packs/ genderandmigration#:~:text=Migration%20can%20lead%20to%20 a%20greater%20degree%20of,bac k%20to%20their%20country%20 of%20origin%20as%20remittances.

11. Pettinger, T. (2019). Pros and cons of Immigration. Economics Help. Retrieved from https://www.economicshelp.org/blog/152453/ economics/pros-and-cons-of-immigration/.

Rosacea Patients Continue To Be at an Increased Risk of Anxiety and Depression

Angela Kazmierczak[1], Neha Saroya[1,2], Muzammil Syed[1], and Austin Mardon[1,3]

Affiliations

1. Antarctic Institute of Canada, #103 11919 82 St NW, Edmonton AB, T5B 2W4, Canada

2. Department of Kinesiology, McMaster University, Hamilton, ON, Canada

3. Department of Psychiatry & John Dossetor Health Ethics Centre, University of Alberta, 2J2.00 WC Mackenzie Health Sciences Centre, 8440 112 St NW, Edmonton AB, T6G 2R7, Canada.

Dermatological researchers say that rosacea patients remain at an increased risk of anxiety and depression and often avoid social situations. The researchers, Monika Heisig and Adam Reich, from the Department of Cosmetology and Dermatology in Poland, assessed the main findings of 13 studies focusing on rosacea and psychological health. The studies analyzed dated from 2005 until 2018 and now serves as a warning as to how the mental impact of the disease, unlike other skin conditions, is still underestimated by clinicians today, despite its debilitating and severe nature (Heisig & Reich, 2018, p. 103).

The studies examined participants from around the world and included samples sizes as high as 13.9 million or 4 million participants to as low as 7 or 17 patients (Table 1).

One of the major findings, noted in 12 of the 13 studies, was increased anxiety and depression amongst both men and women suffering from rosacea (Table 1). Many studies also found that patients with severe rosacea, which is typically found in males, exhibited a stronger fear

34

of blushing, higher cases of social phobia, and elevated stress levels in comparison to non-patients or patients with milder forms of the disease (pp. 104106).

On the other hand, women, who often suffer from the disease more frequently than men, reported lower levels of social anxiety and depression, though their anxiety scores surpassed those of the healthy control subjects (p. 104). According to the studies, the condition's severity is what influences the degree of social anxiety and depression experienced (p. 104).

The belief is that a weakened self-perception is impacting how patients interact with themselves and others, as mentioned in the study (p. 104). According to the researchers, "Facial appearance plays an enormous role in our self-esteem and interactions with other people. Thus, it is not difficult to understand why rosacea patients frequently experience the fear of social judgement" (p. 104).

Heisig and Reich add, "The psychosocial effect of rosacea can be severe and debilitating and lead to social anxiety and depression" (p. 106).

Upon evaluating today's method of treatment and the overwhelming number of studies indicating an increase in total anxiety, Heisig and Reich are convinced that the "negative impact [rosacea has] on psychosocial well-being and on patients' overall quality of life" is being overlooked and left untreated in many doctor offices (p. 104).

Heisig and Reich suggest, "psychological factors, such as stress and anxiety, may even aggravate flushing in rosacea, leading to a vicious cycle. It further supports the importance of considering not only clinical presentation but also the psychological status of patients with rosacea" (p. 104).

In contrast to our current method of treatment, that involves the fixation on the exterior symptoms, Heisig and Reich apply a holistic approach. They believe while diminishing blushing, redness, or other exterior symptoms remains key to eradicating social anxiety and depression in patients, mental wellness also plays a role (p. 104). The researchers hold the notion that improving the mental health of patients and addressing any emotional triggers leads to fewer flare-ups and an improved quality of life.

For patients struggling or in need of additional assistance, the researchers recommend cognitive-behavioural therapy (CBT) be offered as that supplementary treatment, as multiple studies have found it, to an extent, successful in reducing social anxiety (p.106). As confirmed in a number of the articles analyzed, therapy may diminish rosacea remissions, as social interactions often trigger the fears of blushing and flushing (p. 106).

To obtain a better understanding of the condition, popular Youtuber and nutritionist, Naz Ahmed, described the emotional turmoil of her rosacea diagnosis in an interview. "[Upon diagnosis], I decided to stay home most of the time and to not expose my skin to the sun. Also, because I had lost my confidence, I preferred to stay home until I had figured it out" (personal communication, October 24, 2020). As often directed by dermatologists, patients are told to wear sunblock and to avoid sun exposure during 11am to 3pm, as the sun emits its strongest ultraviolet (UV) rays at this time (Canadian Dermatology Association, 2020, p. 1).

Vivian Veroba, an esthetician and rosacea patient in her mid-fifties, disclosed her experience with the inflammatory condition as well. "Even an older woman like myself is affected with low self-esteem because of it. I didn't even want to go out in public some days, especially if I didn't have my coverup" she said. "It's a devastating diagnosis. A quick Google search even says there isn't a cure" (personal communication, October 24, 2020).

Heisig and Reich stress that "the problem of stigmatization and psychosocial distress in rosacea is still underestimated . . . And, it is important for clinicians to acknowledge the psychological impact of this disease to stimulate them to introduce more comprehensive treatment" (p. 106).

Amy Kucherawy, a registered psychologist who practices cognitive-behavioral therapy in Athabasca, Alberta, offered advice on how to rise above the social qualms inflected by rosacea. She said, "It's totally understandable that a rosacea patient would have a total increased self-judgement and to be noticing of self-perception," but "we are always our own worst enemy, and it's important to be your own best friend. Never judge yourself in a way a best friend wouldn't."

In consideration of how uncontrollable rosacea can be, as it can become triggered in social settings, by certain foods, or even by a shift in the environment, Kucherawy advises, "while we do not have control over the rosacea, we can control how we think of ourselves. It's important to be kind to yourself."

As a final note, she said, "Rosacea may be uncontrollable, but we can control how we respond."

There is still much to learn about rosacea and the psychological impact of the condition. As stated in the study, there is unfortunately little known about rosacea, the inflammatory condition, when compared to other skin problems, such as, psoriasis, acne, vitiligo, or atopic dermatitis (p. 104). Researchers hope that in the coming years that narrative will change and more will be understood about the condition.

If you experience any prolonged redness, it is recommended you consult a doctor in order to help treat the redness or to keep it from progressing. The researchers disclose that additional data is required on the topic at large.

References

Canadian Dermatology Association. (2021). Sun Safety for Every Day. https://dermatology.ca/public-patients/sun-protection/sun-safety-every-day/

Heisig, M., Reich, A. (2018). Psychosocial aspects of rosacea with a focus on anxiety and depression. Clinical, Cosmetic and Investigational Dermatology, 11, 103-107. doi: https://dx.doi.org/10.2147%2FCCID.S126850

Mutations in the spike protein of SARS-CoV-2 and their impact on vaccine efficacy

Ananna Bhadra Arna[1], Vivek Kannan[1], Daivat Bhavsar[1], Tian Jian Gao[1], Jasrita Singh[1], Dr. Austin Mardon[1,2]

Affiliations:

1. Antarctic Institute of Canada (AIC), Alberta, Canada
2. John Dossetor Health Ethics Centre, University of Alberta, Edmonton, Alberta, Canada

The zoonotic virus known as severe acute respiratory syndrome coronavirus 2 (SARSCoV-2) uses unstable genetic blueprints, RNA, for inheritance. Susceptibility to high mutations thus becomes an important aspect of the life-cycle of any RNA viruses. It remains questionable whether such genome variability will have any impact on prospective vaccines, particularly if any of the drug targets are directly affected. This paper highlights a few of the mutations most popular over the media and analyzes their capability to undermine the efficacy of the vaccines on hand.

Mutation can occur during repeated events of viral replication. Not all SARS-CoV-2 mutants are dangerous to the host. While some mutations may disable the virus, or weaken its aggression, others may introduce new phenotypes that are favourable for its transmittance. Whether a variant is threatening to the host is determined by various aspects of the viral lifecycle: survival, virulence, and transmission. A variant boosting any of these three processes, alone or in combination, will increase the severity of the virus. For a virus to persist, it must find a compatible host cell for replication. Viral entry for SARS-CoV-2 is an important process that depends largely on the recognition of the host surface receptor Angiotensin I-converting enzyme 2 (ACE2) by the spike (S) transmembrane protein on SARS-CoV-2 (1). Spike protein has two subunits: S1 containing the receptor-binding domain and S2 domain responsible for membrane fusion (2). Any mutation in the domains of spike protein may affect their potential to recognize, infect, and persist into the host cell. Theoretically, a mutation in spike protein should make the virus more infectious and transmissive. 4000 mutations discovered in spike proteins have been discovered across the variants to date (3, 4). However, not all are widespread enough to be worth discussing.

A high frequency of mutations in the Wuhan-1 strain has been reported since its emergence. These mutations can help better suit against the host defence system or as a simple consequence of genetic drift. Most of such mutations are situated at the spike protein. Spike surface glycoprotein mutation D614G has become quite a dominant variant with its presence in 74% of all sequences published up until June 2020 (5). Coevolution of the G614 mutation along with several other synonymous and non-synonymous mutations, varying across strains, elevates the pathogenicity of SARS-CoV-2 (6). The polymorphism from D614 to G614 changes the viral phenotype by boosting its infectivity in the upper respiratory tract and increasing its susceptibility to antibody neutralization. Although the D614 mutation is rather harmless given its high efficacy in antisera-neutralization, this might still pose a threat to monoclonal antibody treatment with its epitope location on the mutated receptor-binding domain (7).

In the summer of 2020, a SARS-CoV-2 outbreak at Dutch mink farms led to viral spillover from minks to the close-contact farmers - which, in turn, resulted in large community transmissions. Among the cases reported, a unique mutant was identified. This new variant with a combination of mutations has been termed as "Cluster 5" (ΔFVI-spike) with four genetic changes with three mutations and one deletion in the spike protein (Y453F, I692V, M1229I, and 69-70deltaHV) that were not previously observed in the strain already circulating. Y453F is a frequent mutation at the receptor-binding domain of spike protein, which was found in the infected mink. The others are deletion of histidine and valine at positions 69-70 of an amino acid chain (69-70deltaHV), a furin cleavage site mutation I692V, and M1229I in its transmembrane domain (8, 9). Y453F, although very prevalent in Dutch mink outbreak sequences, is not an essential mutation in infected minks. From minks to humans, the binding position of spike protein's RBD on an ACE2 host receptor is different – which results in the emergence of Y453F to be compatible with the change at the binding site. This mutation further increases the binding affinity to human ACE2 receptors, thereby increasing viral genome introduction and infectivity in an intercellular manner (9).

The mink-related variants although have not been reported with any enhanced transmissibility, they circulate rapidly in the minks and human population nearby (8). Farmed mink mutations have been reported from nine different countries: Denmark, the Netherlands, Spain, Sweden, Italy, Lithuania, Greece, the United States, and Canada (9,10). It is not often that infected animals can transmit back to humans, and minks are one of the exceptions (11). Minx acts as a reservoir for SARS-CoV-2, and the animal can participate in the spread of more interhuman outbreaks, as well as spillover to other close-contact compatible host species. Furthermore, the viral spread among inter and intraspecies may lead to dangerous genetic modifications as the virus continues to evolve and evade the immune system's defences. Cluster5 itself has not spread widely, however; its genetic modifications (particularly Y453F) have been frequently detected in many samples of mink-related transmissions. It is the impact of these variants on antigenicity that raises the most concern thus far. Research suggests that the mutations do not threaten the efficacy of vaccines yet, however, if not contained now, it may lead to a superstrain that can potentially evade vaccine-induced immunity (12).

B.1.1.7 is a new variant first detected in the UK in mid-September and has been found across the border ever since. This variant has been correlated with a sudden infection spike in the UK however, more research is required to understand whether any causative relationship exists in the new variant to be responsible for high transmissibility. The biggest concern is the rapid spread of the new variant compared to the preceding SARS-CoV-2 virus going around (3). This lineage contains 14 nonsynonymous mutations and 3 deletions. Among the mutations, N501Y and P681H mutations are both situated at the receptor-binding domain. The same deletion at position 69-70 found in mink-related mutations is also observed here (13). The 501Y mutation has already been reported to be present in many other countries like Australia and South Africa (14) without the accompanying mutations found in the new UK variant. Position 501 is part of the binding loop that comes in close contact with the ACE2 receptor and further stabilizes other positions through hydrogen bonding and other interactions. This increases the binding affinity of the mutant (15). While the severity of

40

the variant is still unknown, it can be concluded to possess a higher selective advantage over other variants through evolution by natural selection- proved by their high infection rates. High infectivity however does not necessarily make the variant dangerous as the derived disease does not have severe symptoms (16). To understand what the higher selective advantage entails in terms of phenotypic advantages, more research must be conducted.

A constant fear amongst the vaccine developers is the emergence of a super virus, with mutations unique from the trial strain, that might underplay the efficacy of vaccines being developed. The vaccines at hand are the mRNA-based vaccines: BNT162 by Pfizer and mRNA1273 by Moderna. They both work by introducing mRNA into the body by transportation through lipid nanoparticles. These mRNA strands translate into modified SARS-CoV-2 spike proteins or their components that are recognized as foreign particles and induce an immune response. The mRNA does not integrate into the host genome and is rapidly degraded after inoculation (17). Being an RNA virus capable of zoonotic transmission, SARS-CoV-2 undergoes mutations through its unlimited replication and spillovers. Since most of the reported mutations are observed in the spike protein, particularly in the receptor-binding domain, the undermining of vaccine efficacy is suspected. If the changes in spike protein are too drastic, they may evade the immune system. No mutation has yet escaped the vaccine as they have multiple targets that compensate for the small viral modifications. With more mutations, the vaccine may need to be modified to better suit the target strains as observed with the seasonal flu vaccines (3). However, that does not take away the possibility of the development of a supervirus that is capable of failing vaccines in the future. Until herd immunity is achieved through vaccines, this neverending viral evolution has the potency to sabotage vaccine efficacy (18).

Sources

1. Zumla, A., Chan, J., Azhar, E. *et al.* Coronaviruses — drug discovery and therapeutic options.
Nat Rev Drug Discov 15, 327–347 (2016). https://doi.org/10.1038/nrd.2015.37

2. Belouzard, S., Millet, J. K., Licitra, B. N., & Whittaker, G. R. (2012). Mechanisms of coronavirus cell entry mediated by the viral spike protein. *Viruses*, *4*(6), 1011–1033. https://doi.org/10.3390/v4061011

3. Wise J. Covid-19: New coronavirus variant is identified in UK. BMJ. 2020 Dec 16;371:m4857. doi: 10.1136/bmj.m4857. PMID: 33328153.

4. Junxian Ou, Zhonghua Zhou, *et al*. Emergence of RBD mutations from circulating SARSCoV-2 strains with enhanced structural stability and higher human ACE2 receptor affinity of the spike protein [J]. bioRxiv. 2020.

5. Plante, J. A. et al. Spike mutation D614G alters SARS-CoV-2 fitness. Nature https://doi.org/10.1038/s41586-020- 2895-3 (2020).).

6. **doi:** http://dx.doi.org/10.2471/BLT.20.253591

7. Kannan, S.R., Spratt, A.N., Quinn, T.P. et al. Infectivity of SARS-CoV-2: there Is Something More than D614G?. J Neuroimmune Pharmacol (2020). https://doiorg.cyber.usask.ca/10.1007/s11481-020-09954-3

8. European Centre for Disease Prevention and Control. Detection of new SARS-CoV-2 variants related to mink - 12 November 2020. ECDC: Stockholm; 2020.

9. https://files.ssi.dk/Mink-cluster-5-short-report_AFO2 (preprint)

10. The Mink Pandemic Just Keeps Getting Worse - The Atlantic

11. Why infected mink are raising fears about a vaccine-resistant strain of Covid-19 (newstatesman.com)

12. https://www-nature-com.cyber.usask.ca/articles/d41586-020-03218-z

13. Rambaut A, Loman N, Pybus O, et al. Preliminary genomic characterisation of an emergent SARS-CoV-2 lineage in the UK defined by a novel set of spike mutations. 18 Dec 2020. https://virological.org/t/preliminary-genomic-characterisation-of-an-emer-

gent-sars-cov-2lineage-in-the-uk-defined-by-a-novel-set-of-spike-mutations/563.

14. World Health Organization. SARS-CoV-2 variant—United Kingdom. 21 Dec 2020. https://www.who.int/csr/don/21-december-2020-sars-cov2-variant-united-kingdom/en.

15. https://www.krisp.org.za/manuscripts/MEDRXIV-2020-248640v1-de_Oliveira.pdf

16. New and Emerging Respiratory Virus Threats Advisory Group. NERVTAG meeting on SARS-CoV-2 variant under investigation VUI-202012/01. 18 Dec 2020.
https://khub.net/documents/135939561/338928724/SARS-CoV-2+variant+under+investigation%2C+meeting+minutes.pdf/962e866b-161f-2fd5-103032b6ab467896?t=1608491166921.

17. Moderna COVID vaccine becomes second to get US authorization (usask.ca)

18. Does Virus Mutation Affect Vaccine Efficacy? - Bloomberg

Are COVID-19 Vaccines Solutions for Everyone?

Ananna Bhadra Arna[1], Vivek Kannan[1], Daivat Bhavsar[1], Tian Jian Gao[1], Jasrita Singh[1], Dr. Austin Mardon[1,2]

Affiliations:

1. Antarctic Institute of Canada (AIC), Alberta, Canada
2. John Dossetor Health Ethics Centre, University of Alberta, Edmonton, Alberta, Canada

After being authorized by federal health agencies for emergency usage, many batches of Pfizers and Moderna vaccines have rolled out in different countries across the globe. Although this comes as 'a light at the end of the tunnel' for the ongoing pandemic, considerations need to be made for their impact on people under special groups. As the population is being immunized based on their potential risk of infection, women experiencing pregnancy are doubtful of being injected as no official research has been conducted on vaccine benefits and side-effects in such patients. In more than 300 vaccine trials worldwide, pregnant women were universally excluded - who make up approximately 131 million of the entire population per year worldwide. Such research becomes increasingly relevant as more expectant mothers that test positive for COVID-19 end up hospitalized and ICU admitted. The infection has also been associated with higher rates of cesareans (48.3%), preterm births with less than 37 weeks gestation (21.8%), and perinatal mortalities - to name only a few out of many more complexities. Even though mother-to-child or vertical transmission is deemed very rare,a recent study found about 1.9% of infants to be COVID-positive following birth from an infected mother. However, it is debatable whether the infection is being transmitted to the child before or after birth and requires more investigation. Higher rates of fetal harm, miscarriage, and mortality of both the mother and the neonate have been associated with previous pandemics, such as SARS-CoV and MERS-CoV infection. As they both belong to the same family of RNA virus as SARS-CoV-2, these outcomes are likely present in COVID-19 patients. Furthermore, the possible burden of the infection on the fetal development during gestation period remains unanswered. As body experiences change during pregnancy, the immune system weakens, letting other opportunistic infections to invade. According to WHO, pregnant ladies with pre-existing conditions such as hypertension, diabetes, overweight, older age, etc. are more susceptible to the infection. Thus, their safety becomes absolutely necessary as pregnancy leads to severe symptoms upon infection.

While the risk of transmission in women experiencing normal pregnancy is similar to that of the non-pregnant population, the outcomes they face, once infected, renders them extremely vulnerable. Greater admission of pregnant COVID-19 patients at the ICU indicates the urgency of introducing some preventive measures so that two lives can be saved at once. Although vaccines are out, the expectant moms should be put higher in the priority list with a physician's consent as immunization during pregnancy will have trifold benefit - defending the mother, the fetus, as well as the infant against the virus. However, both Pfizer and Moderna have excluded pregnant women from their trial population in fear of bringing unknown harm to the fetus and the mother (i.e. congenital defects as observed in the thalidomide tragedy). There are still some recordings of women who became pregnant during the trial and were discontinued from the study as they had a strict no-pregnancy policy for the participants. Those who had this exposure early in their pregnancy did not report any adverse event so far. While this sounds reassuring, it does not take away the anxiety that comes with vaccination in the pregnant women. With so many uncertainties, the choice of being immunized becomes rather subjective. It is recommended to have a discussion with her physician to reach a decision with shared consents by both the pregnant recipient and her physician after considering the benefits and possible side-effects associated. In particular, the pre-existing health conditions should be taken into account when deciding to be immunized. During consideration, the candidates should be thoroughly analyzed to choose the vaccine that is more appropriate during pregnancy based on the technology involved (i.e. avoiding vaccines containing live virus for pregnant women). If the expecting mom decides to go through vaccination, then they should come higher in the priority ladder. This is to say, they should be vaccinated at a fast priority basis once they consent for vaccination. Further research is required to understand whether any passive protection is shared with the fetus in the uterus so that the vaccination can be received at an appropriate trimester for maximal protection. To make an informed decision, the pros and cons of vaccine candidates should be thoroughly understood by analyzing its safety in the mother, fetus and infant, and this inclusion of pregnant participants in medical trials is indispensable.

From Parthenon to Pandemic: The Intrinsic Discrimination in Societies Against Individuals with Disabilities Throughout History and how the COVID-19 Response Highlights Such Disparities

Amal Rizvi[1,3], Austin Mardon, CM Ph.D[1,2]

[1]Antarctic Institute of Canada, Edmonton, AB, Canada
[2]University of Alberta, Edmonton, AB, Canada
[3]Western University, London, ON, Canada

ABSTRACT

The dominant culture in the west originated from the notion that individuals living with all types of disability are unable to contribute to society. Thus, the norms, policies, and attitudes that govern societies today often intrinsically marginalize the disabled, and do not account for their unique needs. Though this pattern has been evident throughout history, the COVID-19 pandemic illustrates various disparities and gaps that prevent disabled individuals from accessing the care and information they require.

INTRODUCTION

Since December of 2019, severe acute respiratory syndrome coronavirus 2, more concisely known as SARS-CoV-2, has propelled the global community into a pandemic that many are now deeming "the new normal"[1].As the world simultaneously reels and recovers from the crippling effects of the coronavirus pandemic, the global situation presents the opportunity for disability studies – an extensive sphere of influence that intertwines the deeply personal experiences of individuals who identify with disability with the academic areas of research, theory, and advocacy[2] – to be examined under a novel, culture-defining lens.

Oftentimes, the field of disability studies is a drastically overlooked aspect of academia in today's climate. However, the rampant spread and worldwide social impact of the virus highlights the disparities and inadequacies in the way the dominant culture regards disability-inclusive healthcare accessibility and emergency planning, as well as Western society's response to infected persons with disability and impairments. The disparities and inadequacies faced by the disabled community are not recent developments in society as a result of the global pandemic; rather, they are part of a historical paradigm that has polluted social institutions, as well as cultural and social norms for centuries.

The Centers for Disease Control and Prevention defines disability as a physical or mental condition that limits one's ability to perform certain activities or tasks, thus limiting one's ability to interact with their surroundings[3]. In the context of this review, it is important to streamline this definition into three main ideas:

Firstly, although impairments, the inability or limited ability of a normal human function is a primary aspect of disability, it is important to reinforce that disability becomes apparent in a social context when impairments clash with one's physical or social environment[2]. Second, to understand that disability comes to light in a social context when impairments clash with the environment requires the acknowledgement that every individual's environment is highly unique, and is a product of numerous factors, including, but not limited to, one's culture, personal relationships, physical and geographical structures as well as social institutions, such as government and schooling[2]. Thirdly, disability, in the context of this paper, is examined in a sociocultural and thus, disablism, or the discriminatory, oppressive behaviours and attitudes towards disabled people, is examined as a social pathology[2], rather than a dilemma faced by any given individual. As a whole, disability is a deeply personal experience, while simultaneously being the subject of politicization and public discourse, often in a negative light. Disability studies not only analyzes, but also challenges this populist opinion and the social stigmas it comes with.

The societal response to the coronavirus has been, in short, incomplete. According to the United Nations, people who experience disabilities are disproportionately affected COVID-19, often due to being at a greater risk of contracting the disease or having limited access to support services that they require[4]. Ironically enough, although people living

48

with disabilities are at an increased risk of morbidity and mortality, and make up over 1 billion people worldwide , current strategies to eradicate and prevent contraction of the coronavirus have been anything but disability inclusive. It is these shortcomings in current healthcare protocols implemented in the worldwide COVID-19 response that, in fact, further marginalizes people with disabilities, and builds on the pre-existing disparities and inadequacies they encounter under normal circumstances, let alone in these unprecedented times.

In this paper, we examine how individuals living with disabilities have been historically marginalized and segregated from the common culture, and how societal culture was intentionally formed without addressing the needs or contributions of the disabled. We also highlight three of the major disparities that the disabled community faces in the western healthcare system during the coronavirus pandemic, as a result of being intrinsically segregated from the common culture. We address how future disability-inclusive responses and emergency planning to large-scale global events, like the COVID-19 pandemic must be changed to adequately bridge the large gaps that withhold individuals who identify as disabled from accessing the care and support they need in unprecedented times.

A Brief History of the Marginalization of the Disabled in Societies

Historically, the idea of living with disability has often been accompanied with an apparently unfortunate and bleak perception of a disabled individuals future and quality of life; although the scope of disability is broad, and encompasses, but is not limited to, physical, developmental, intellectual and motor aspects of one's life, the entire notion of disability has been historically and falsely labelled as a hallmark of societal and personal failure.

Throughout human history, disabled individuals have not only been marginalized and rejected as they are by modern cultures today, but also wrongfully persecuted, and even murdered. In Sparta, for instance, ancient Greeks were required by law to commit infanticide when children were born with an evident disability or deformity[6]. In medieval Europe, Malleus Maleficarum, the most renown work written about witchcraft, stated that disabled children were Satan's spawn , and consequently resulted in the torture and murder of individuals with[7] psychiatric disorders and disabilities; this further reinforced that disability was a physical manifestation of divine punishment.

The marginalization of the disabled continued into the industrial era, in which these individuals were placed in institutions, and denied the chance to become productive members of society[6]. Although the industrial revolution is frequently characterized as a time of innovation, development, and a pivotal point that gave rise to urbanized societies, it is also a time in which disabled individuals were often unknowingly forced into custody, because they were deemed unfit for the discipline, rigor, and skill needed for factory work. Disabled individuals were though to be roadblocks to a modern world – too slow and inflexible for the hustle and bustle of this time period. With the rise of social Darwinism in the late 1800s, many Eugenicists feared that the disabled were genetically linked to crime, poverty and unemployment and rallied to sterilize mentally and physical disabled individuals[6].

In this way, the laws, norms and dominant culture that govern society were never built with disabled persons taken into consideration. The economy and society built from this era and onwards required disabled individuals to be tucked away and institutionalized, often for the majority of their lives, while the economic, social, and cultural norms were developed without the presence, let alone their input.

Even in today's seemingly modern society, the negative connotation associated with words like "disabled" or "impaired" exists to an alarming degree. Globally, it is often acceptable for doctors to allow newborns with obstructive impairments to die if the parents consent[6]. Throughout childhood, disabled children are four times as likely to be victims of violence than those without disabilities, with children with cognitive disabilities being the most vulnerable individuals[8]. Similarly, adults with disabilities are 1.5 times more likely to be victims of violence[8] than non-disabled adults . The probability of violence towards disabled individuals increases with the rise of ignorance and stigma, which, as of today, is rampant in schools, workplaces and even disabled persons' own households. Whether it be in the form of an inability to obtain employment due to ableism in the workplace, or parents exhibiting blatant favoritism towards their non-disabled children, the marginalization that disabled individuals experience regularly is a prevalent problem in society; it is ingrained in the common culture for people to celebrate ability, and simultaneously disregard disability as something less meaningful in life.

DISPARITIES AND GAPS IN DISABILITY INCLUSIVE HEALTHCARE DURING THE CURRENT CORONAVIRUS PANDEMIC

The Presence of Intrinsic Biases in COVID-19 Response Planning Mimics Historic Discrimination Against the Disabled

The association of disability with societal failure shapes the perception of disability in the context of large-scale global events, such as the COVID-19 pandemic. Policy makers, healthcare professionals, as well as the myriad of other individuals responsible for developing, executing and implementing emergency planning and prevention measures to mitigate the spread of coronavirus can possess the same intrinsic biases that have existed throughout history; ableism is a social prejudice that is learned, adopted and reinforced by individuals in the dominant culture and, consequently, it is a paradigm that is indirectly or directly present in COVID-19 response procedures. The following sections examine how medical rationing and triage assessments, as well as preventative measures implemented in western society, such as social distancing and isolation, are aspects of the COVID-19 response that have been largely ableist; adequate alternatives or solutions have not been curated in order for disabled members of society to partake in these measures safely and comfortably. Further, we address how poverty, a major social determinant of health, was not sufficiently accounted for in regards to emergency planning for the disabled community.

Medical Rationing and Triage Assessments for COVID-19 Treatment is Intrinsically Discriminatory

Medical rationing, the process by which resources, treatment, and access to the healthcare system is allocated to individuals, is often necessary as a result of economic limitations within the healthcare system, or varied cost and benefit to individuals' wellbeing[9]. Throughout the process of rationing, however, individuals with disabilities are often further marginalized and neglected in favour of individuals deemed as more important or more able to contribute. In fact, people with disabilities often remain invisible or unrecognized during times of crisis and, as a result, risk management procedures and preventative measures often do not account for the correct, and often personalized, care that these individuals need[10].

Amidst the COVID-19 pandemic, there has also been an emphasis on the distribution of treatment via a triage framework that typically favours patients who have the highest likelihood of survival; consequently, this type of assessment neglects people with disabilities, as it is a common notion in society that conventionally able-bodied individuals have a higher likelihood of survival[11]. In this way, the healthcare response for the current coronavirus pandemic has been majorly ableist in the sense that medical rationing for ventilators, other treatment measures, and even spaces in hospitals has continuously excluded those who are most vulnerable, which includes the disabled[11] due to preconceived notions about their potential quality of life[12]. This type of triage framework has, unfortunately, been adopted as a consistent process of ethical decision making in relation to the COVID-19 pandemic and implements a one-size-fits-all approach that can be detrimental to the health and wellness of disability communities in the west[12].

Advocates for disability rights argue that the protocols for triage assessments must be modified from a focus on an individual's long term survival – which is intrinsically discriminatory and biased as it places a disability identity on an individual and does not solely account for their current condition in respect to coronavirus – and instead, assess near-term survivability[13]. It is argued that this type of survivability is independent from the disability identity[13], which, when placed upon an individual by a physician rather than the disabled individual themself, is rooted in ableism.

Measures Implemented to Prevent the Spread of Coronavirus Do Not Consider the Needs and Lifestyles of the Disabled Community

Disabled individuals who require in-home or in-person care from nurses or other caretakers may find themselves struggling to meet their needs on their own, as one of the largest components of the COVID-19 response in the west has been to limit physical interaction [14]. The increased reliance of digital medicine and E-health during the pandemic[15] does not necessarily address the needs of disabled individuals, especially for those who cannot easily access the required

52

technology on their own. For instance, individuals who do not know how to use computers, individuals in rural areas without stable internet access, or individuals living in restrictive home environments in which their family members prevent them from accessing a cell phone are only a few examples of members of the disability community who will not be able to receive adequate care remotely; this can have detrimental impacts on their health, mental wellbeing, and life quality. The lack of persistent or regular care results in the exacerbation of pre-existing health conditions and co-morbidities[14]. Further, many individuals with disabilities require regular check-ups or screenings, to ensure that pre-existing conditions and future diagnoses do not become fatal[14]; the disruption to these services strengthens the barrier that withholds the disabled community from accessing the level of care they may need to thrive in a society that already holds them back.

The promotion of social distancing and isolation, although deemed essential to prevent the spread of the virus, can have detrimental effects on mental and physical health of individuals in the disability community. Cognitive or mental disabilities can progressively worsen in isolation, as individuals lack social interaction and mental stimulation for an extended period of time[16]. Many disabled individuals were already vulnerable to the deleterious mental health impacts of prolonged social isolation; the current pandemic has only exacerbated this vulnerability [17], and with no definite idea of an end in sight, emotional difficulties, anxiety, depression and added stress can consume individuals' lives during quarantine. Frequent calls or text messages with their caretakers and aids, as well as family members can potentially mitigate some of the effects of social isolation. However, healthcare practitioners cannot depend on their patient's families and friends to maintain frequent communication with a disabled relative, and with the elevated level of stress caretakers and other aids, such as nurses, have faced during the pandemic[18], facilitating this additional level of communication on their behalf is not only difficult and unpredictable, but also unfeasible. The technology barrier that some disabled individuals experience can also hinder frequent communication and aggravate negative mental health and depressive thoughts.

Poverty and lower socioeconomic status (SES) has a more prevalent effect during the COVID-19 pandemic on the disabled community[14], which is already disproportionately affected by poverty[19] in the west, than in the past. Basic necessities that disabled individuals require and may have scarcely been able to afford prior to the pandemic, are now inaccessible in the midst of the financial crisis that exacerbated the difficulties faced by impoverished disabled individuals[20].

Grocery shopping, for instance, may have already been troublesome and stressful for disabled individuals living in poverty, but now, with the increased risk of infection that could be fatal to their health, disabled individuals may be unable to purchase groceries entirely[14]. Online grocery shopping and delivery services, though often proposed as an alternative, are not only difficult for some disabled individuals to access and navigate but are often more costly[20].

Individuals with low SES are also more likely to live in smaller, more crowded apartments or shared housing complexes in which the spread of coronavirus is more rampant, and social distancing measures are difficult to implement[20]. This increased risk of infection[17] increases mortality for already vulnerable disabled individuals, who can often suffer from co-morbidities, and exacerbates negative emotions pertaining to stress, anxiety and paranoia, originating from a crippling and very reasonable fear of contracting the virus[14].

In parts of the west, namely the United States, disabled individuals of lower SES may be unable to seek the treatment they need in the event that they contract the virus. With the average household containing a disabled adult requiring 28 percent more income than the average household without[21], the extra cost of living with disability is high enough to severely hinder disabled adults below the poverty line from comfortably integrating in society in normal circumstances. With hospital stays, treatment methods, and even vaccines having a price tag attached, the cost of both contracting, as well as preventing the spread and mortality of the virus in the disabled community can add up to more than an impoverished, disabled individual can afford.

CONCLUSION

Curating a Disability-Inclusive Response: Future Considerations

It is not enough to simply acknowledge the presence or needs of people living with disability or impairments – rather, a crucial facet of curating a disability-inclusive response to large-scale global events like the COVID-19 pandemic is to offer people with disabilities the opportunities to participate in the creation and implementation of emergency planning[10]. This ensures that people living with disabilities can maintain dignity and feel that their needs are adequately met[10] .Conventionally able-bodied individual cannot speak over disabled individuals; the assumption that emergency planning procedures designed by able-bodied individuals in regards to the current pandemic are sufficient for disabled individuals is blatantly ignorant and reinforces how the common culture of society is built and continues to be built with the absence of people with disabilities. Further, it is not the role of disabled individuals alone to rally to be recognized as equal members of the society in which they live. Though the forthcoming of a universal disability culture opposes the idea that disability is a personal weakness and indication of social failure[2], able-bodied individuals must adopt this idea into the common culture, especially in global crises, such as the coronavirus pandemic.

Emphasis must also be placed in individuals having the potential to receive unbiased care and treatment to recover from COVID-19[10]. Access to healthcare, treatments and preventative measures should not consider aspects of one's personal identity, such as race, religion, employment, and disability. The triage approach physicians use to assess for further treatments in regards to the coronavirus must thus be unbiased and exclude these factors from consideration, even if these factors allegedly contribute to long-term survivability[13]. In fact, physicians should move away from the assessment of long-term survivability of an individual in regards to coronavirus treatment procedures, as long-term survivability is vague and difficult to predict on short notice[13]. For example, an individual presenting a physical disability may be immediately rejected for coronavirus treatment due to the intrinsically ableist and discriminatory perception that this individual will not live as long as an able-bodied individual. Meanwhile, an able-bodied individual with an underlying and undiagnosed respiratory condition may be approved for treatment immediately.

From the era of the Parthenon in ancient Greece[6] to the coronavirus pandemic, the negative stigmas and stereotypes associated with individuals living with disabilities have been ingrained in the social structure, institutions and even the schemas of individuals responsible for maintaining order in society during unprecedented times. In the healthcare system, the role of physicians is to provide treatment to those who require it, but intrinsic physician biases that reinforce negative stereotypes about people living with disabilities can cloud this process with bias, and potentially shorten a disabled individual's lifespan more than their own disability ever could.

ACKNOWLEDGMENT

We acknowledge the Antarctic Institute of Canada for supporting this research, as well as the #RisingYouth initiative on behalf of the Canada Service Corps, TakingITGlobal and the Government of Canada.

1. Wu, Yi-Chi, Chen Ching-Sung, and Chan Yu-Jiun. 2020. *The outbreak of COVID-19: An overview*. J Chin Med Assoc. 83(3): 217–220.
2. Goodley, Dan. 2017. *Disability Studies: An Interdisciplinary Introduction* .
3. Centers for Disease Control and Prevention. 2020. *Disability and Health Overview.*
4. United Nations. 2020. *Policy Brief: A Disability Inclusive Response to COVID-19.*
5. Groce, Nora Ellen. 2018. *Global disability: an emerging issue.* Lancet Glob Health. 6(7):E724–E725.
6. Barnes, Colin. 1991. *Disabled People in Britain and Discrimination: A case for anti-discrimination legislation.*
7. Haas, LF. 1992. *Neurological stamp: Johanne Weyer (Weir).* J Neurol Neurosurg Psychiatry . 55(5): 346.

8. World Health Organization. *Violence against adults and children with disabilities.*

9. Chen, Bo and McNamara, Donna Marie. 2020. *Disability Discrimination , Medical Rationing   and COVID-19.* Asian Bioeth Rev. 1–8.

10. World Health Organization. 2013. *Guidance Notes on Disability and Emergency Risk   Management for Health.*

11. Andrews, Erin E., Ayers, Kara B., Brown, Kathleen S., Dunn, Dana S., and Pilarski, Carrie R. 2020. *No Body is Expendable: Medical Rationing and Disability Justice During the COVID-19 Pandemic.* Am Psychol. http://dx.doi.org/10.1037/amp0000709

12. Goggin, Gerard and Ellis, Katie. 2020. *Disability, communication, and life itself in the COVID-19 pandemic.* Health Sociol Rev. 29(2): 168–176.

13. Solomon, Mildred Z., Wynia, Matthew K., Gostin, Lawrence O. 2020. *Covid-19 Crisis   Triage – Optimizing Health Outcomes and Disability Rights.* N Engl J Med. 383. 10.1056/NEJMp2008300

14. Lund, Emily M., Forber-Pratt, Anjali J., Wilson, Catherine, and Mona, Linda R. 2020. *The COVID-19 Pandemic, Stress, and Trauma in the Disability Community: A Call to Action.* Rehabil Psychol. 65(4). 313–322.

15. Thulesius, Hans. 2020. *Increased importance of digital medicine and eHealth during the   Covid-19 pandemic.* Scand J Prim Health Care. 38(2): 105–106.

16. Holt-Lunstad, Julie., Smith, Timothy B., and Layton, J. Bradley. 2010. *Social relationships   and mortality risk: A meta-analytic review.* PloS Med. https://doi.org/10.1371/journal.pmed.1000316

17. Olsen, Jason. 2018. *Socially disabled: the fight disabled people face against loneliness and stress.* Disabil Soc. 33(7): 1160–1164.

18. Krystal, John H. 2020. *Responding to the hidden pandemic for healthcare workers: stress.* Nat Med. 26. https://doi.org/10.1038/s41591-020-0878-4

19. Erikson, William, Lee, Camille G., and von Schrader, Sarah. 2019. *2017Disability Status Report United States.*

20. Armitage, Richard and Nellums, Laura B. 2020. *The COVID-19 response must be disability inclusive.* Lancet Public Health. https://doi.org/10.1016/S2468-2667(20)30076-1

21. National Disability Institute. *The Extra Costs of Living with Disability in the U.S. – Resetting the Policy Table.*

Coronavirus: A Blessing in Disguise?

Ashley Meelu, Jasrita Singh, Austin Mardon

The COVID-19 pandemic has changed the world completely. Nonetheless, it is hard to say that the change has been negative for every living organism. While the pandemic has taken a devastating toll on humankind, the wildlife population of Canada is essentially thriving. Likewise, due to the lockdown, many individuals have started working from home, resulting in fewer cars on the road and hence, significantly lower pollution levels. This reduction has also been imposed by the decrease in vacationing and planning of road trips. As the air is now much clearer in the once more contaminated areas of the GTA, many animals have had the opportunity to venture out further to make their habitats. Moreover, it has allowed for the life expectancy of many wildlife animals to increase, as they are breathing in and are surrounded by less polluted air.

As a result of the tighter regulations being placed in the light of the pandemic, many activities of hunting and shooting have also been ceased. Paired with the reduction in driving, there has been a consequent decrease in the death toll of many animals. Among the luckier are those animals that are ordinarily run over by cars or used as targets for hunting. Activities such as poaching of exotic animals have also been placed at an all-time low. This is because it is harder to gain equipment and vehicles without fear of virus transmittance. The pandemic has caused many officials to tighten the current regulations placed on the treatment of wildlife. Thus, the wildlife populations are now at a lesser danger from humans.

Another positive side effect of the lockdown protocol is the importance that the general population is placing on their hygiene. With this special consideration but putting in cleanliness, public areas have become much cleaner and devoid of litter. As animals such as birds are often found travelling through public areas, they are now less prone to ingesting litter or being embroiled in unsanitary situations. Additionally, the refrain of people from travelling into public areas has permitted wildlife to disperse among greater areas of the communities. With this, animals can find better areas for habitation as well as spreading their species has allowed for better population distribution.

Though the pandemic has created a difficult new world for humankind, it has been increasingly apparent that it has its benefits. Fundamentally, no system or situation can be eternally or completely ideal for any population- whether it be humans or animals. Individuals have gained many healthy practices in light of the virus. These include but are not limited to consideration of cleanliness, sneezing and coughing into one's arm and maintaining a safe distance. These practices not only keep humans safe from each other and their environment but also shelters animals. Undeniably, the wildlife population has long suffered at the hands of humankind. It has taken our complete lockdown to allow animals to live freely and safely. This sad reality can hopefully be changed, however, by the same- if not somewhat adapted- consideration of the needs of animals in the future.

***

Ashley Meelu, BSc is an undergraduate student at the University of Waterloo with a background in Biomedical Sciences.

Jasrita Singh, BHSc is an undergraduate student at McMaster Universtiywith a background in Biochemistry, Biomedical Discovery and Commercialization.

Austin Albert Mardon, CM, FRSC is an adjunct professor in the Faculty of Medicine and Dentistry at the University of Alberta, an Order of Canada member, and Fellow of the Royal Society of Canada.

The Impact of COVID-19 on Small Businesses and the Tourism Industry

Lajendon Jeyakumar, Jasrita Singh, Austin Mardon

The current pandemic has made some strides to improve the government's responsibility for its people, however, how much damage has been done for countries and their economies? According to the World Health Organization (WHO), as of August 17, 2020, there have been 21, 756, 357 confirmed cases of COVID-19 worldwide, and 771, 635 lives lost. Along with the numerous tragic deaths, the global economy is steadily declining due to a myriad of unprecedented factors including the restrictions on travel and the loss of restaurant income. This has led to many businesses having to permanently close their doors and a countless number of jobs lost.

According to Restaurants Canada, the large decline in revenue has resulted in 36% of restaurant and other food business owners to believe that they will only restore profitability in 12 to 18 months. This is not only a red flag for the country's economy, but has also created a steep hole for many restaurant owners to come out of. Currently, 290,000 jobs are being offered, as opposed to approximately 1.2 million people employed in the food sector before the pandemic. The food sector has long remained the fourth largest private sector in Canada to help provide jobs. Losing this industry, including restaurants, can have detrimental consequences on Canada's national employment rate. With ripple effects visible in various other industries such as agriculture, these losses may leave a lasting dent on the Canadian economy.

Similarly, in the study conducted by Alexander Bartik and colleagues, 43% of the 5819 small businesses surveyed in the United States of America, were found to be temporarily closed, resulting in a 40% reduction of employees. This remains to be the ground reality after the implementation of the CARES act by the U.S. government, which helps provide financial relief for affected small businesses. Similarly, the Canadian government has implemented a Canadian Emergency Commercial Rent Assistance (CECRA), which particularly provides financial assistance for small businesses and restaurants undergoing financial hardships and problems with rent. To stabilize the employment outcomes, the Canadian government has introduced the Canada Emergency Wage Subsidy (CEWS) to cover a portion of employee's wages, but this is dependent on the restaurant's financial standing.

With a loss of employment and a declining economy, many countries are on the brink of crumbling. The introduction of travel bans and personal reservations to travelling has also limited the income many countries obtain through tourism. According to the World Tourism Organization, it is stated that "international travel could lose 850 million to 1.1 billion international tourists, which can lead to a loss of US $910 billion to 1.2 trillion dollars in revenue from tourism". Countries heavily reliant on the tourism and hospitality industries such as Mexico, Iceland, Jamaica, Croatia, and Thailand continue to experience a large deficit.

Despite the loss of businesses and lack of travelling, all countries are implementing the necessary steps to give stability to their respective countries. Hoping that things change for the better and we all stand united, that is all we can look forward to achieving together.

***

Lajendon Jeyakumar is an undergraduate student pursuing a Bachelors of Life Sciences at University of Toronto, with a major in Psychology, and minors associated with Public Law and Biology.

Jasrita Singh is an undergraduate student pursuing a Bachelors of Health Sciences at McMaster University, with a specialization in Biomedical Discovery and Commercialization.

Austin Albert Mardon, CM, FRSC (University of Alberta) is an adjunct professor in the Faculty of Medicine and Dentistry, an Order of Canada member, and Fellow of the Royal Society of Canada.

Decentralizing Health-Care Sites to Better Serve People Experiencing Homelessness

Viveka Pimenta

In the midst of the public health emergency posed by COVID-19, it is important to remember the lived reality of people battling another public health crisis: homelessness. The obstacles homeless people face every day are already near impossible to deal with and are exacerbated during a pandemic which blocks access to public services and threatens the alreadyfragile safety of people without housing.

Despite homeless people being more vulnerable to chronic and acute disease, very few of them have a primary care physician. They rely upon walk-in clinics, community health centres, and emergency rooms for health care, or simply do not have a regular source of health care. Many lack an OHIP card, which is another barrier to healthcare access in times of need.

The lack of barrier-free access to health care is increasing overall expenses through the use of walk-in clinics, emergency rooms, and dependency on an OHIP card. A 2012 paper reports that hospitalization of a homeless person is almost five times as expensive as the hospitalization of a housed person. It is expensive to rely upon emergency services and given that homeless people have few other ways of coping with their increased risk for illnesses, it appears to be their only choice.

In 2018, a collaborative project was launched between the Toronto Central Local Health Integration Network and the City of Toronto in efforts to improve access to health care for residents of shelters. Five new shelters were unveiled with over 300 beds reserved for residents with complex illnesses, and a committee of health care providers, residents and operators of shelters originated to advise on the process of increasing health care access for shelter residents. Alex Zsager, co-chair of the network's Citizens' Panel, said that as a former shelter resident, he could attest to the feelings of despair that set in rapidly, especially alongside other health conditions. Zsager stated, "Wait times for such services could be as long as 6 months to a year and this is unacceptable. By providing access to proper health care and services to residents in our shelters you would see a drop in emergency visits and better health for all."

Homeless people are far more vulnerable to COVID-19 infection and are far more likely to develop severe illness. The homeless in Ontario, according to the Canadian Medical Association Journal Open, face odds of being twenty times more likely to be hospitalized for COVID-19 than the general housed population, in addition to being ten times more likely to require intensive care and five times more likely to die within three weeks of contracting the infection.

People experiencing homelessness are unable to comply with public health mandates for at-home isolation, and their inconstant access to hygiene, healthcare, and basic necessities pose obstacles for remaining safe from the virus.

In addition to this, homeless people are more prone to complex comorbidities that can prolong and worsen COVID-19, as they experience higher rates of mental illnesses, cardiovascular and pulmonary diseases, diabetes, and high blood pressure compared with the public.

Many homeless people are afraid to catch the virus and will abstain from living in crowded shelters or accessing regular services to support mental health and addiction rehabilitation. This makes it difficult to ensure immunization via the two vaccine doses delivered weeks apart. People experiencing homelessness need a trusting relationship with their healthcare providers, and this need is not being met by public health officials. Having health care professionals who treat the population on a daily basis and become recognizable to them can alleviate some of the strain and make it more likely for homeless people to willingly seek out consistent healthcare, follow up on appointments, and become more trusting of a healthcare system that supports them fully.

The sector is overwhelmed trying to respond to a public health emergency that it is not designed for.

Toronto, with the largest shelter system in Canada, has been using a three-tier approach responding to the pandemic in an effort to protect the vulnerable without overwhelming the health care system. The first tier, prevention, has involved further funding to all of the city's shelters, training for service providers, distributing personal protective equipment to the homelessness sector, physical distancing, and opening temporary residences for the homeless. It involves risk stratification by identifying homeless individuals who face the greatest risk in the pandemic and moving them into temporary housing. Street outreach includes the increase of cleaning and safety for the outdoor encampments, screening homeless people for symptoms while providing referrals and hygiene kits and distributing information on encampment-specific strategies to social distance.

The second tier, mitigation, implements a standard screening for COVID-19 at all points of entry, transports homeless clients to assessment centres so they can receive testing, prioritizes testing in shelters, and provides space for a 14-day isolation program with healthcare for people who screen positive for COVID-19.

The third tier, recovery, has prioritized providing recovery and isolation spaces for homeless individuals whose illnesses are not severe enough for the hospital, which reduces the strain on emergency health care services while also reducing risk of viral spread throughout the shelters. An additional facet of the recovery tier is rapid rehousing, leveraging investments to provide permanent housing, and ensuring housing stability to minimize the increased vulnerability faced by the homeless and prevent critical illness.

These efforts can be improved further to expand healthcare access to these vulnerable populations in uncertain times. The current lag in vaccine shipments to the country should be viewed as a much-needed opportunity to strengthen the programs in place for vaccine delivery by decentralizing vaccination sites and educating the community to combat any lingering hesitancy to be vaccinated.

In early February, a report was released by Ontario's COVID-19 science advisory table regarding Israel's vaccine delivery system, which worked speedily and strategically to immunize the population as evident by the steady drop in cases and hospitalizations. At the report's time of release, around 80 per cent of adults over 50 were vaccinated in Israel.

Key to the success of vaccine delivery was the use of pop-up and drive-through centres as a means of decentralizing care away from hospitals and pharmacies to increase efficiency and accessibility. Establishing vaccine sites where nurses and physicians could administer doses within communities and remote locations allowed for swift vaccination of the majority of the population. It was also key to create a line of contact between the health care system and the leaders of every individual community, which allowed a sophisticated treatment plan to be curated for each location.

They ensured that vaccine supply could safely be delivered to all locations, not just hospitals. Vaccine shipments from Pfizer are typically accompanied by a technical requirement wherein 975 doses must be transported in a large freezer tray that makes it difficult to transport and distribute in small, remote settings. Israel worked with Pfizer to repack the pallets in each tray into small insulated 300-dose boxes to alleviate this challenge and allow for local shipping.

If Ontario contacts vaccine manufacturers and repackages the vaccine shipments, it could allow for near seamless vaccine distribution to smaller at-risk communities, retirement homes, and remote locations, promoting the decentralization of vaccine sites away from hospitals.

Israel required people to fill out a questionnaire which determined their eligibility for a vaccine. Once it was signed, it was considered to be implied consent, which allowed expedited vaccination without a formal written consent process, while all further documentation happened digitally.

Implementing these decentralized vaccine sites in Ontario could even out vaccine rates across communities and allow for tailoring of vaccine delivery to marginalized communities. Implied consent would make this standard of healthcare more accessible to homeless individuals while simultaneously expediting the process for all populations.

In addition, Israel utilised community-based health care professionals to deliver the COVID-19 vaccine so quickly. This can be attributed to the trusting relationships that these practitioners, paramedics, and nurses have already formed with community members. Doing this in Ontario could remove one further obstacle preventing at-risk individuals from receiving protection from the COVID-19 virus.

However, as Canada is a much larger country, decentralization won't be nearly as quick. Each province is working on its own vaccine rollout system, and until the next shipment of shots, this extra time should be utilized to develop these decentralized clinics and booking systems.

Currently in Toronto, nine new city-run vaccination sites are on schedule to be operative by early April 2021. They constitute a network of immunization clinics throughout Toronto that will dispense 120 000 doses of vaccine weekly. Once the vaccine is resupplied, Toronto plans to immunize the masses through these new city-operated clinics, mobile clinics, hospitals, and eventually pharmacies and family practitioners. Many local health centres are using this time to provide information to communities demonstrating vaccine hesitancy stemming from long-term trust issues with the health care system. There is a need for the government to fortify these efforts by ensuring every community has access to resources that will prepare them when the next vaccine rollout arrives.

These efforts to educate will also keep outbreaks in high-risk populations at bay as they will be more aware of their risk and methods for following public health guidelines that are tailored to their individual situations.

In the future, continued usage and improvement of the strategies that have been developed during this pandemic could improve healthcare access for the homeless in uncertain times. Decentralizing sites of healthcare away from the hospital can alleviate the strain faced by the system, ensuring a streamlined approach for future public health emergencies to minimize risk to our most vulnerable populations.

Increasing location coverage allows more efficient healthcare delivery to more of the population, ensuring preparedness for public health emergencies. Mobile clinics bring healthcare to marginalized and homeless communities in a climate that simultaneously stigmatizes these vulnerable individuals while expecting them to come forward and rely upon hospitals and clinics. As mobile clinics supply socks, underwear, hygiene products, drug recovery kits, and Naloxone kits, they should be a permanent health service in major cities to protect vulnerable populations from their increased adversity.

By increasing barrier-free access to healthcare through mobile clinics, street nurses, shelter staff physicians, and overall, more sites of health care that do not rely upon hospital resources, we can alleviate some of the suffering for homeless people and also save money by reducing the strain on these systems. This money should be put towards housing for homeless people and preventing vulnerable populations from becoming homeless.

Decentralized sites of healthcare could also allow for hospitals to prioritize homeless people who truly require hospital care, streamlining the transfer from shelters. It is imperative to place priority on individuals in true need, especially since poor healthcare exacerbates the struggles of homelessness. More sites of healthcare allow for more familiarity of healthcare personnel with local homeless populations, so some of the barriers, like lack of government ID, can be eliminated.

The lack of trust for the healthcare system experienced by marginalized communities and homeless people is perpetuated by standardized, inflexible care that does not carve out a space where these individuals fit. People with lived experiences in homelessness do not experience patient-centred care in a manner that is universally expected, due to the inherent biases of healthcare providers towards patients with a higher socioeconomic status.

Increasing sites of healthcare and establishing an Equity-Oriented Healthcare framework that is tailored to each community could reduce some of the long-standing mistrust experienced by vulnerable people by treating them with compassionate, personalized, and accessible healthcare.

What other services could be decentralized from hospitals to increase accessibility for all of the population and improve the health of Toronto's homeless population?

***

Viveka Pimenta, BMSc (University of Western Ontario) is an undergraduate student studying biochemistry and pathology. She researches and writes interdisciplinary articles with the Antarctic Institute of Canada.

References

Choi, Y., Stall, N. M., Maltsev, A., Bell, C. M., Bogoch, I. I., Brosh, T., Evans, G. A., Grill, A., Hopkins, J., Kaplan, D. M., McGeer, A., Moran-Gilad, J., Nowak, D., Presseau, J., Salmon, A., Schwartz, B., & Juni, P. (2021). Lessons Learned from Israel's Vaccine Rollout. Science Briefs. https://doi.org/10.47326/ocsat.2021.02.09.1.0

City of Toronto COVID-19 Response for People Experiencing Homelessness. (2020, October 14). City of Toronto. https://www.toronto.ca/news/city-of-toronto-covid-19-response-forpeople-experiencing-homelessness/

Fitzpatrick, S. (2018, January). Toronto Central Local Health Integration Network (LHIN). Www.torontocentrallhin.on.ca. http://www.torontocentrallhin.on.ca/MessagefromtheCEO/HomelessSheltersandHealthCa re.aspx

Gaetz, S. (2012). The real cost of homelessness: Can we save money by doing the right thing? | The Homeless Hub. Www.homelesshub.ca; Toronto: Canadian Homelessness Research Network Press. https://www.homelesshub.ca/costofhomelessness

Gulliver, T. (2014). How Can We Improve Healthcare Access for the Homeless? | The Homeless Hub. Www.homelesshub.ca. https://www.homelesshub.ca/resource/how-can-weimprove-healthcare-access-homeless

Krugel, L., & The Canadian Press. (2021, February 3). "Imperative" COVID-19 vaccines prioritized for homeless, shelter staff: advocates. Global News. https://globalnews.ca/news/7616388/covid-canada-homeless-vaccine/

Leary, B. (n.d.). Unhoused in Toronto: The delivery and experience of hospital healthcare services for homeless people. Retrieved February 28, 2021, from https://www.homelesshub.ca/sites/default/files/attachments/Unhoused%20in%20Toronto%20-%20Bill%20O%27Leary.pdf

Pelley, L. (2021, February 6). Canada faces lull before COVID-19 vaccine shipments ramp up — and it's a race to prepare | CBC News. CBC.https://www.cbc.ca/news/health/covid-19vaccine-shipments-lull-1.5901457

Purkey, E., & MacKenzie, M. (2019). Experience of healthcare among the homeless and vulnerably housed a qualitative study: opportunities for equity-oriented health care. International Journal for Equity in Health, 18(1). https://doi.org/10.1186/s12939-019-1004-4

Rodrigues, G. (2021, February 19). City of Toronto to open 9 COVID-19 vaccine clinics to administer shots. Global News. https://globalnews.ca/news/7650033/toronto-covid19vaccine-clinic-locations-coronavirus/

Sheikh, M. (2021, January 23). Mobile health clinic to support Toronto's most marginalized communities - CityNews Toronto. Toronto.citynews.ca.https://toronto.citynews.ca/2021/01/23/mobile-health-clinic-to-support-torontos-most-marginalized-communities/

Understanding the Pathogenesis of COVID-19, including Symptoms and Treatments.

N. Kang[1], J. Singh[1], A. Mardon[2]

Affiliations:

1. Department of Biochemistry and Biomedical Sciences, McMaster University, Hamilton, Ontario, Canada

2. John Dossetor Health Ethics Centre, University of Alberta, Edmonton, Alberta, Canada

Introduction

In December 2019, a virus now named as the severe acute respiratory syndrome coronavirus 2 (SARS-CoV-2) created a series of atypical respiratory diseases originating in Wuhan, Hubei Province, China[1]. This virus created a disease known as COVID-19. It is believed that the outbreak originally started via a zoonotic transmission associated with exposure to the wet animal market in Wuhan City[2]. The pandemic of COVID-19 spread rapidly, creating great global public health concern. It has impacted a large number of people worldwide, being reported in approximately 200 countries and territories[2]. Due to the person-to-person transmission of COVID-19, infected patients were led into isolation while being administered a variety of treatments. Many extensie measurements have been introduced in order to reduce person-toperson transmission, particularly to protect susceptible populations such as the elderly[2]. The management of COVID-19 has been incredibly difficult due to numerous reasons, including the high infectivity of the virus, large asymptomatic populations, and lack of ineffective antivirals and vaccines[3]. The knowledge related to the pathobiology of this disease is quite limited and in order to reach acceptable conclusions, previous knowledge related to SARS-CoV must be utilized[1].

Symptoms

On average, the symptoms of COVID-19 appear after an incubation period of approximately 5.2 days[4]. Some of the most common symptoms include fever, cough, fatigue headaches, diarrhoea, haemoptysis, dyspnoea and lymphopenia[5]. There have been some general similarities in the symptoms between COVID-19 and the previous betacoronavirus. However, COVID-19 has shown some unique clinical features, which includes the focus on lower airways, explaining the upper respiratory tract symptoms[3]. These symptoms include rhinorrhoea, sneezing and sore throat[3] (see Figure 1). To add on, previous coronaviruses did not exhibit intestinal symptoms such as diarrhoea, while a larger percentage of patients infected with COVID-19 did[3]. Some similarities between COVID-19 and the earlier betacoronavirus include fever, dry cough, dyspnea and bilateral ground-glass opacities as seen on chest CT scans[3]. The period from the onset of COVID-19 symptoms to the patient's death ranged from 6 to 41 days, with a median of 14 days[5]. This time period is dependent on various factors, including the patient's immune system and their age[3]. This period was shorter for patients over the age of 70 years old compared to those that were younger[5]. By looking at the cells that are likely infected, COVID-19 can be divided into three phases that correspond to the different clinical stages of the disease.

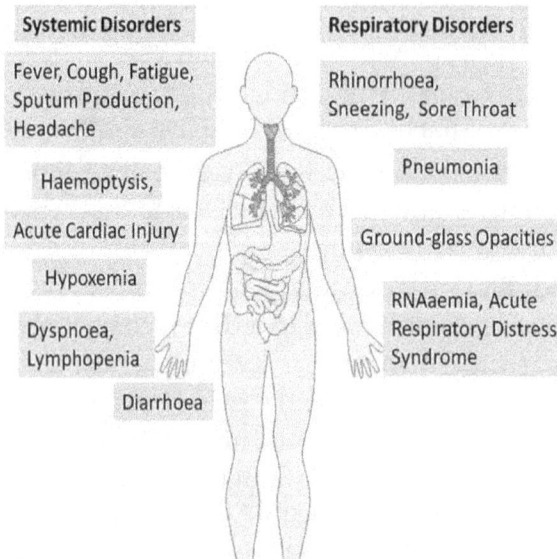

Systemic Disorders

Fever, Cough, Fatigue, Sputum Production, Headache

Haemoptysis,

Acute Cardiac Injury

Hypoxemia

Dyspnoea, Lymphopenia

Diarrhoea

Respiratory Disorders

Rhinorrhoea, Sneezing, Sore Throat

Pneumonia

Ground-glass Opacities

RNAaemia, Acute Respiratory Distress Syndrome

Figure 1: This exhibits the systemic and respiratory disorders caused by COVID-19 infection. A chest CT scan revealed abnormal features such as RNAaemia, acute respiratory distress syndrome, acute cardiac injury, and incidence of ground-glass opacities, which led to death in some cases[3]. In some patients, the multiple peripheral ground-glass opacities were seen in subpleural regions of both lungs, including an immune response and leading to increased inflammation[3]. These stages are based on the assumptions that the viral entry by SARS-CoV2 will be the same as SARS-CoV, as we do not know if there is an alternative receptor for SARS-CoV[3].

Stage 1

Stage 1, also known as the asymptomatic state, consists of the initial one to two days on infection. During this stage, it is likely that the inhaled virus binds to epithelial cells that are found in the nasal cavity[6]. This is where it begins to start replicating. ACE2 is known to be the main receptor for both SARS-CoV2 and SARS-CoV[7]. Based on in vitro experiments, it is believed that the ciliated cells are the primary cells infected in the conducting airways[8]. However, other experimental data does not support this hypothesis as single-cell RNA indicated low levels of ACE3 expression in the conducting airway cells, along with no cell type preference[8]. During this stage, there is a very limited innate immune response, and only local propagation of the virus[1]. A nasal swab may be used to detect the virus at this stage and although patients may not be experiencing many symptoms, these patients remain infectious[1]. Nasal swabs are believed to be more sensitive than throat swabs when it comes to detecting the virus[1]. The RT-PCR value for the viral RNA can be useful in determining the viral load, the infectivity and the clinical course[1]. This value could possibly allow super spreaders to be detected, helping to reduce the spread. However, in order to use RT-PCR cycle values, the sample collection must be standardized.

Stage 2

The next stage of the virus consists of the upper airway and conducting airway response. The virus now begins to propagate and migrate down the respiratory tract, including the conducting airways, which then triggers a more innate immune response[1]. A nasal swab or sputum at this stage would yield the virus along with early markers of the innate immune response[1]. By examining the level of CXCL 10, or possibly other innate response cytokines, one may be able to predict the subsequent clinical course[9]. CXCL 10 is an interferon responsive gene, known to be useful as a disease marker in SARS and influenza[10]. The viral defected epithelial cells are a major source of beta and lambda interferons[11]. By examining the host innate immune response, it may be possible to improve predictions on the subsequent course of the disease[1]. For approximately 80% of the infected patients, the disease will be restricted to the upper and conducting airways, while presenting mild symptoms[12]. These individuals do not need aggressive monitoring and may be monitored at home with conservative symptomatic therapy.

Stage 3

Unfortunately, approximately 20% of the infected patients will go past stage 2 and enter stage 3 of the disease[12]. These infected patients will develop pulmonary infiltrates and some may develop a very serious form of the disease[1]. Currently, the estimates of the fatality rate are about 2% but this varies according to the age of the patient[12]. These estimates may become better defined once the prevalence of mild and asymptomatic cases is better understood. During this stage, the virus will now reach the gas exchange units of the lung and will begin to infect the alveolar type II cells[13] (see Figure 2). These infected alveolar units are likely to be peripheral and subpleural[14]. This is a similarity between SARS-CoV and influenza, as they both preferentially infect type II cells rather than type I cells[15]. As the virus propagates within the type II cells, large numbers of viral particles are being released, causing the cells to undergo apoptosis and die[10]. This results in a self-replicating pulmonary toxin, as the viral particles from the infected type II cells infect the adjacent cells[1]. Due to this rapid spread of viral particles, large areas of the lungs lose most of their type II cells, triggering the secondary pathway for epithelial regeneration[1]. Type II cells are of high importance, as they are typically the precursor cell for type I cells[16]. The pathological result of COVID-19 is diffuse alveolar damage with fibrin rich hyaline membranes and a few multinucleated giant cells[17]. The abnormal wound healing may result in scarring and fibrosis[18]. Recovery from these results may need a very vigorous innate and acquired immune response along with epithelial regeneration. At this stage, it may seem useful to administrate epithelial growth actors such as KFG, but this may be detrimental due to increased viral load by producing more ACE2 expressing cells, as seen with influenza[19]. The elderly are at a higher risk due to their diminished immune response and reduced capacity to repair the damaged epithelium[1]. The elderly are particularly at risk due to their reduced mucociliary clearance, allowing the virus to easily spread to the gas exchange units of the lungs[20].

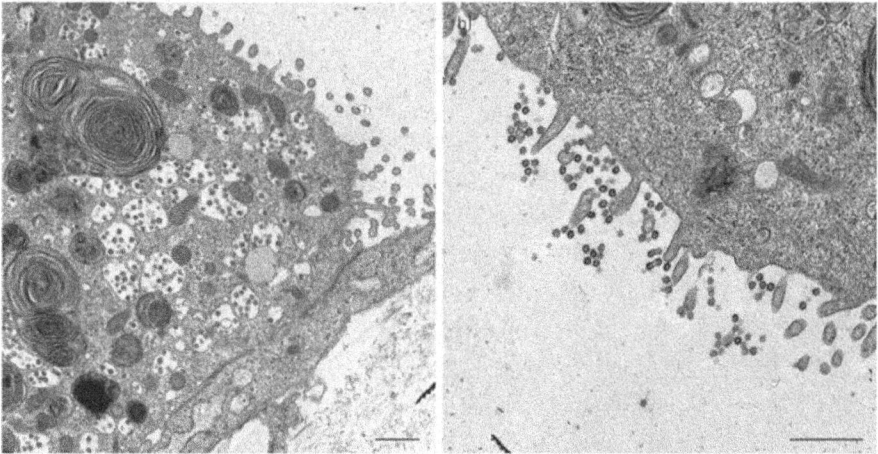

Figure 2: Human alveolar type II cells were isolated, cultured *in vitro* and then infected with SARS-CoV[10]. The viral particles are seen in the double membrane vesicles in the type II cells. They can be seen along the apical microvilli[10].

Transmission

It is suggested that there is likely a zoonotic origin of the COVID-19 as it is believed to have originated in the wet animal market in Wuhan City[3]. There were a large number of infected people that were exposed to this market. The search for a reservoir host or intermediate carriers are ongoing, but this would help to explain how the infection spread to humans. There have been two species of snakes that are being investigated as possible reservoirs of the virus[3]. However, the experimental results have been inconclusive, and consistently display mammals and birds as coronavirus reservoirs[21]. It is likely that mammals are the most likely link between COVID-19 and humans as suggested using genomic sequence analysis[22]. It has also been observed that person-to-person transmission is the likely route utilized by COVID-19, which is evident in individuals and families who did not visit the wet animal market in Wuhan City[3]. Transmission between individuals occurs through direct contact or through droplets that are spread by an infected individual through coughing and sneezing[23]. A study was conducted on pregnant women in order to determine whether there is transmission from mother to child, and it showed that infected women in their third trimester did not transmit the virus to their children[24]. However, all these women had a cesarean section performed, meaning further studies on pregnant women that include vaginal birth are required[14].

Pathogenesis

There have been an increasing number of infected individuals and fatalities, particularly in the epidemic region of China[3]. When patients infected with COVID-19 were examined, it was discovered that they possess higher leukocyte numbers, abnormal respiratory findings and an increased level of plasma pro-inflammatory cytokines[3]. Many individuals infected with COVID19 presented a cough, coarse breathing sounds from both lungs, a persistent fever and a body temperature averaging at 39.0°C[3]. The COVID-19 infection was confirmed using real-time polymerase chain reaction[25]. The patients also showed leucopenia, having leukocyte counts of 2.91 x 10^9 cells/L, of which 70.0% were neutrophils[3]. To add on to that, the normal range of blood C-reactive proteins is 0-10 mg/L but infected individuals had an average value of 16.16 mg/L[3]. In these patients high erythrocyte sedimentation rate and D-dimer was also exhibited[25]. Additionally, it was discovered that infected patients had higher blood levels of cytokines and chemokines, which included IL1-β, IL1RA, IL7, IL8, IL9, IL10, basic FGF2, GCSF, GMCSF, IFNγ, IP10, MCP1, MIP1α, MIP1β, PDGFB, TNFα, and VEGFA[26]. Some of the infected patients had a more severe case, requiring the need of an intensive care unit, and these patients exhibited high levels of proinflammatory cytokines including IL2, IL7, IL10, GCSF, IP10, MCP1, MIP1α, and TNFα[26]. These proinflammatory cytokines are believed to promote disease severity. The main findings in terms of pathogenesis of COVID-19 infection as a respiratory system targeting virus include RNAaemia, acute cardiac injury, ground-glass opacities and severe pneumonia[26].

Phylogenetic Analysis

COVID-19 is classified as a β CoV of group 2B by the World Health Organisation (WHO)[27]. Ten genomic sequences of COVID-19 were taken from nine patients and this exhibited a 99.98%[22] sequence identity. Another similar experiment using five patients demonstrated a 99.899.9%[28] nucleotide identity in the isolates. This study also revealed a new beta-CoV strain. By analyzing the genetic sequence of COVID-19 it was discovered that it showed more than an 80% similarity to

SARS-CoV and 50% to MERS-CoV, which both originate in bats[28]. Using this evidence, it is clear that COVID-19 belongs to the genus betacoronavirus, which is the same genus as SARS-CoV, and infects humans, bats and wild animals[29]. The COVID-19 is classified under the orthocoronavirinae subfamily, which consists of seven members of the coronavirus family that have the ability to infect humans[29]. This virus forms a clade within the subgenus sarbecovirus. Although COVID-19 is similar to SARS-CoV, there are significant differences that permits it to be considered a new betacoronavirus that infects humans[3]. Along with its similarity to SARS-CoV and MERS-CoV, there is more evidence that supports the fact that COVID-19 is of bat origin. This includes the high degree of homology of the ACE2 receptors found in various animal species, making it a possibility that these animal species may act as possible intermediate hosts[6]. Continuing on, this virus has a single intact open reading frame on gene 8, which is a strong indicator of bat-origin CoVs[6]. The amino acid sequence of the receptor-binding domain is similar to that of SARS-CoV, which indicates that these two viruses may use the same receptor[28].

Possible Treatments

As mentioned earlier, COVID-19 spreads person-to-person, which led to the isolation of infected patients. These isolated patients were administered a variety of treatments but there are no specific antiviral drugs or vaccines that have proven successful at the moment[3]. The current option is to utilise broad-spectrum antiviral drugs, such as Nucleoside analogues and HIV-protease inhibitors which work to attenuate the viral infection until a better treatment option is available[30]. A study including 75 patients included the administration of existing antiviral drugs. This included twice a day oral administration of 75 mg oseltamivir, 500 mg lopinavir, 500 mg ritonavir and the intravenous administration of 0.25 g ganciclovir for 3-14 days[31]. Another study showed that broadspectrum antiviral remdesivir and chloroquine showed a significant improvement in infected patients in vitro[3]. These antiviral compounds have been previously used in human patients, allowing them to be considered to be used to treat COVID-19.[32] There are other compounds that are being developed for the treatment of COVID-19 infection, which includes EIDD-2801, a compound that has shown promising results against seasonal and pandemic influenza virus infections[33]. As more time is needed until specific therapeutics become available for the treatment of

COVID-19, it is important to consider more broad-spectrum antivirals. Some of these broadspectrum antivirals include Lopinavir/Ritonavir, Neuraminidase inhibitors, peptide (EK1), RNA synthesis inhibitors[3]. However, it is clear that there is more research required urgently needed to identify drugs for treatment COVID-19. In order to do this, there is an urgent need to create an animal model to replicate this disease, allowing the development of pre- and post-exposure prophylaxis. There is ongoing research, with several groups of scientists working to develop a nonhuman primate model, which would allow them to study the COVID-19 infection and test potential vaccines while better understanding the virus-host interactions[3].

Future Ideas

As there is no specific antiviral drug or vaccine for COVID-19, other extensive measures are required. This includes extreme measures to reduce the person-to-person transmission of COVID19, which is absolutely essential to control the current outbreak. These efforts to reduce transmission would be especially applied to susceptible populations, such as children, health care providers, and elderly people. It has been observed that the early death cases of the infection occurred primarily in elderly people[34]. This may be due to the fact that the older individuals have a weaker immune system, which permits faster progression of the virus[1]. In order to prevent the spread of the virus, public services should provide decontaminating reagents for cleaning hands in a routine manner. In addition to that, fecal and urine samples may potentially drive as an alternative route for transmission, so special attention should be applied when dealing with these objects along with other wet and contaminated objects[34]. Many countries have implemented various control measures, including travel screenings and mandatory 14-day isolation in order to control further spread of the virus[23]. The virus must be studied further, including any epidemiological changes. This includes studying new potential routes of transmission, adaptation, evolution and virus spread between humans and possible animals and reservoirs. There are still many questions that need to be answered, including who to test and how many tests need to be administered. There is also further research required in numerous cases, such as the fact that there have been very few pediatric cases, which may be due to lack of testing or a true lack of infection. By asking more questions and working to answer them, this will provide a framework to answer more specific questions, allowing more extensive public health measures to be implemented.

Conclusion

The current COVID-19 pandemic is a live issue that is affecting people all over the world. The number of death tolls continues to rise and many countries are being forced to go into lockdown and practice social distancing. It has changed the way the world works, with stricter regulations to prevent further spread of the virus. Currently, the lack of targeted therapies continues to be a significant problem, but broad-spectrum antiviral drugs are currently being administered[35]. There is a lot of ongoing research on various topics related to COVID-19, with the hope that we will be able to understand the virus better. Epidemiological studies have shown that some populations such as the elederly are more susceptible and require special attention while children tend to have milder symptoms[35]. Genetic analysis has allowed the classification of the virus along with finding similarities between COVID-19 and previous viruses[1]. By using information gathered from previous viruses, the three clinical stages of COVID-19 have been established, which allows a better understanding of how the virus enters the body and progresses. The symptoms and likelihood of death can range based on numerous factors, such as the strength of the immune system and the age. Various experiments have demonstrated the impact of the virus on the body, such as the changes observed in alveolar type II cells[10]. It's transmission method is quite clear but it may be possible for it to be transmitted in other ways, such as through fecal matter. It is not clear whether there is an animal reservoir and if it is truly from a bat origin, but further research will confirm these results[3]. Finding an animal model would facilitate the vaccine creation process, as this animal model could be used to better understand the way the virus interacts with the host and would allow vaccines to be tested. Without a specific antiviral drug, the current focus is to reduce the spread of the virus while providing supportive care for infected individuals[2].

Acknowledgements:

The authors would like to sincerely thank Professor Catherine Mardon for her supervision and support during the construction of this paper.

Financial Support:

This paper received $350 in grant funding from the TakingItGlobal charity allocated for the publishing of this article in a professional publication source.

Declarations of Conflict of Interest: None

Data sharing is not applicable to this paper because no new data was created or analyzed.

Author Biographies:

Navneet Kang, BSc (McMaster University) is an undergraduate student with a background in Chemical Biology Jasrita Singh, BHSc (McMaster University) is an undergraduate student with a background in Biochemistry, Biomedical Discovery and Commercialization. Austin Albert Mardon, CM, FRSC (University of Alberta) is an adjunct professor in the Faculty of Medicine and Dentistry, an Order of Canada member, and Fellow of the Royal Society of Canada.

Works Cited

1. Mason RJ. Pathogenesis of COVID-19 from a cell biology perspective. European Respiratory Journal. 2020;55(4). doi:10.1183/13993003.00607-2020
2. Yuki K, Fujiogi M, Koutsogiannaki S. COVID-19 pathophysiology: A review. Clin Immunol. 2020;215:108427. doi:10.1016/j.clim.2020.108427
3. Rothan HA, Byrareddy SN. The epidemiology and pathogenesis of coronavirus disease (COVID-19) outbreak. Journal of Autoimmunity. 2020;109:102433. doi:10.1016/j.jaut.2020.102433
4. Li Q, Guan X, Wu P, et al. Early Transmission Dynamics in Wuhan, China, of Novel Coronavirus–Infected Pneumonia. New England Journal of Medicine. Published online January 29, 2020. doi:10.1056/NEJMoa2001316
5. Wang W, Tang J, Wei F. Updated understanding of the outbreak of 2019 novel coronavirus (2019-nCoV) in Wuhan, China. Journal of Medical Virology. 2020;92(4):441-447.

doi:10.1002/jmv.25689

6.   Wan Y, Shang J, Graham R, Baric RS, Li F. Receptor Recognition by the Novel
     Coronavirus from Wuhan: an Analysis Based on Decade-Long Structural Studies of SARS
     Coronavirus. Journal of Virology. 2020;94(7). doi:10.1128/JVI.00127-20

7.   Hoffmann M, Kleine-Weber H, Schroeder S, et al. SARS-CoV-2 Cell Entry Depends on ACE2 and TMPRSS2 and Is Blocked by a Clinically Proven Protease Inhibitor. Cell. 2020;181(2):271-280.e8. doi:10.1016/j.cell.2020.02.052

8.   Reyfman PA, Walter JM, Joshi N, et al. Single-Cell Transcriptomic Analysis of Human Lung Provides Insights into the Pathobiology of Pulmonary Fibrosis. Am J Respir Crit Care Med. 2018;199(12):1517-1536. doi:10.1164/rccm.201712-2410OC

9.   Tang NL-S, Chan PK-S, Wong C-K, et al. Early Enhanced Expression of InterferonInducible Protein-10 (CXCL-10) and Other Chemokines Predicts Adverse Outcome in Severe Acute Respiratory Syndrome. Clin Chem. 2005;51(12):2333-2340. doi:10.1373/clinchem.2005.054460

10. Qian Z, Travanty EA, Oko L, et al. Innate Immune Response of Human Alveolar Type II Cells Infected with Severe Acute Respiratory Syndrome–Coronavirus. Am J Respir Cell Mol Biol. 2013;48(6):742-748. doi:10.1165/rcmb.2012-0339OC

11. Hancock AS, Stairiker CJ, Boesteanu AC, et al. Transcriptome Analysis of Infected and Bystander Type 2 Alveolar Epithelial Cells during Influenza A Virus Infection Reveals In Vivo Wnt Pathway Downregulation. Journal of Virology. 2018;92(21). doi:10.1128/JVI.01325-18

12. Wu Z, McGoogan JM. Characteristics of and Important Lessons From the Coronavirus Disease 2019 (COVID-19) Outbreak in China: Summary of a Report of 72 314 Cases From the Chinese Center for Disease Control and Prevention. JAMA. 2020;323(13):1239-1242. doi:10.1001/jama.2020.2648

13. Mossel EC, Wang J, Jeffers S, et al. SARS-CoV replicates in primary human alveolar type II cell cultures but not in type I-like cells. Virology. 2008;372(1):127-135. doi:10.1016/j.virol.2007.09.045

14. Wu J, Wu X, Zeng W, et al. Chest CT Findings in Patients

With Coronavirus Disease 2019 and Its Relationship With Clinical Features. Investigative Radiology. 2020;55(5):257–261. doi:10.1097/RLI.0000000000000670

15. Zhang S, Li H, Huang S, You W, Sun H. High-resolution computed tomography features of
17 cases of coronavirus disease 2019 in Sichuan province, China. European Respiratory Journal. 2020;55(4). doi:10.1183/13993003.00334-2020

16. Yee M, Domm W, Gelein R, et al. Alternative Progenitor Lineages Regenerate the Adult Lung Depleted of Alveolar Epithelial Type 2 Cells. Am J Respir Cell Mol Biol. 2016;56(4):453-464. doi:10.1165/rcmb.2016-0150OC

17. Gu J, Korteweg C. Pathology and Pathogenesis of Severe Acute Respiratory Syndrome.
The American Journal of Pathology. 2007;170(4):1136-1147. doi:10.2353/ajpath.2007.061088

18. Xu Z, Shi L, Wang Y, et al. Pathological findings of COVID-19 associated with acute respiratory distress syndrome. The Lancet Respiratory Medicine. 2020;8(4):420-422. doi:10.1016/S2213-2600(20)30076-X

19. Nikolaidis NM, Noel JG, Pitstick LB, et al. Mitogenic stimulation accelerates influenzainduced mortality by increasing susceptibility of alveolar type II cells to infection. PNAS. 2017;114(32):E6613-E6622. doi:10.1073/pnas.1621172114

20. Ho JC, Chan KN, Hu WH, et al. The Effect of Aging on Nasal Mucociliary Clearance, Beat Frequency, and Ultrastructure of Respiratory Cilia. Am J Respir Crit Care Med. 2001;163(4):983-988. doi:10.1164/ajrccm.163.4.9909121

21. Bassetti M, Vena A, Giacobbe DR. The novel Chinese coronavirus (2019-nCoV) infections: Challenges for fighting the storm. European Journal of Clinical Investigation. 2020;50(3):e13209. doi:10.1111/eci.13209

22. Lu R, Zhao X, Li J, et al. Genomic characterisation and epidemiology of 2019 novel coronavirus: implications for virus origins and receptor binding. The Lancet. 2020;395(10224):565-574. doi:10.1016/S0140-6736(20)30251-8

23. Carlos WG, Dela Cruz CS, Cao B, Pasnick S, Jamil S. COVID-19 Disease due to SARSCoV-2 (Novel Coronavirus). Am J Respir Crit Care Med. 2020;201(4):P7-P8. doi:10.1164/

rccm.2014P7

24. Chen H, Guo J, Wang C, et al. Clinical characteristics and intrauterine vertical transmission potential of COVID-19 infection in nine pregnant women: a retrospective review of medical records. The Lancet. 2020;395(10226):809-815. doi:10.1016/S01406736(20)30360-3

25. Lei J, Li J, Li X, Qi X. CT Imaging of the 2019 Novel Coronavirus (2019nCoV)　　　　Pneumonia. Radiology. 2020;295(1):18-18. doi:10.1148/radiol.2020200236

26. Huang C, Wang Y, Li X, et al. Clinical features of patients infected with 2019 novel coronavirus in Wuhan, China. The Lancet. 2020;395(10223):497-506. doi:10.1016/S01406736(20)30183-5

27. Hui DS, Azhar EI, Madani TA, et al. The continuing 2019-nCoV epidemic threat of novel coronaviruses to global health — The latest 2019 novel coronavirus outbreak in Wuhan, China. International Journal of Infectious Diseases. 2020;91:264-266. doi:10.1016/j.ijid.2020.01.009

28. Ren L-L, Wang Y-M, Wu Z-Q, et al. Identification of a novel coronavirus causing severe pneumonia in human: a descriptive study. Chinese Medical Journal. 2020;133(9):1015–1024. doi:10.1097/CM9.0000000000000722

29. Zhu N, Zhang D, Wang W, et al. A Novel Coronavirus from Patients with Pneumonia in China, 2019. New England Journal of Medicine. Published online January 24, 2020. doi:10.1056/NEJMoa2001017

30. Lu H. Drug treatment options for the 2019-new coronavirus (2019-nCoV). BioScience Trends. 2020;14(1):69-71. doi:10.5582/bst.2020.01020

31. Chen N, Zhou M, Dong X, et al. Epidemiological and clinical characteristics of 99 cases of 2019 novel coronavirus pneumonia in Wuhan, China: a descriptive study. The Lancet. 2020;395(10223):507-513. doi:10.1016/S0140-6736(20)30211-7

32. Wang M, Cao R, Zhang L, et al. Remdesivir and chloroquine effectively inhibit the recently emerged novel coronavirus (2019-nCoV) in vitro. Cell Research. 2020;30(3):269-271. doi:10.1038/s41422-020-0282-0

33. Toots M, Yoon J-J, Cox RM, et al. Characterization of orally efficacious influenza drug with high resistance barrier in

ferrets and human airway epithelia. Science Translational Medicine. 2019;11(515). doi:10.1126/scitranslmed.aax5866

34. Lee N, Hui D, Wu A, et al. A Major Outbreak of Severe Acute Respiratory Syndrome in Hong Kong. New England Journal of Medicine. 2003;348(20):1986-1994. doi:10.1056/NEJ-Moa030685

35. Cao W, Li T. COVID-19: towards understanding of pathogenesis. Cell Research. 2020;30(5):367-369. doi:10.1038/s41422-020-0327-4

Meta-Analysis on the Efficacy of Midodrine in the Treatment of Neurocardiogenic Syncope

Eman Zaheer[a], Janani Rajendra[b], Ai Ling Zhu[c], Zain Siddiqui[d], Syed Muhammad Ali Salman[d], Ismaeel Mohammedally[e] and Yang Zhao[f]

Affiliations:

[a]Department of Biological Sciences, University of Toronto, Toronto, Canada; [b]Faculty of Science, University of Western Ontario, London, Canada; [c]Schulich School of Medicine & Dentistry, University of Western Ontario, London, Canada; [d] Faculty of Science, McMaster University, Hamilton, Canada; [e]Faculty of Engineering, Ryerson University, Toronto, Canada; [f]Richmond Hill High School

Abstract

The human body can be triggered by certain stimuli, which causes it to overreact; this condition is known as neurocardiogenic syncope. Symptoms of this disease include a sudden drop-in heart rate and blood pressure, ultimately causing a loss of consciousness. Other forms of treatment, such as the use of β-blockers, have been explored, but these treatments have been shown to be ineffective, and in some cases, may even worsen symptoms. Midodrine, an αadrenergic agonist, has inconclusive evidence of its ability to treat neurocardiogenic syncope. Therefore, using previous clinical studies, this meta-analysis investigated the efficacy and reliability of midodrine for neurocardiogenic syncope. In this analysis, four articles using the variables of midodrine administration and recurrence rates of syncope were identified and the numerical data was processed through RevMan to find that there was a significant difference in syncope recurrence between the midodrine and placebo groups (p = 0.0004). The meta-analysis concludes that there is a strong correlation between administration of midodrine and an improvement in syncope recurrence in patients with neurocardiogenic syncope.

Introduction

Neurocardiogenic syncope, also known as vasovagal syncope, is a condition in which the human body is triggered by certain stimuli and overreacts by influencing the circulatory system (Brignole, 2007). These triggers may include extreme heat or the sight of blood, and it can cause an individual's heart rate and blood pressure to drop suddenly, leading to a brief loss of consciousness (Fogoros, 2020; Brignole, 2007). The mechanism of how neurocardiogenic syncope occurs is still poorly understood, but it is caused by the activation of cardiac C fibers (Chen-Scarabelli & Scarabelli, 2004). Symptoms that occur in presyncope, which is the feeling of lightheadedness before syncope, may include dizziness, nausea, and sweating (Pietrangelo, 2019).

First-line treatments for neurocardiogenic syncope are currently non-pharmacological treatments, such as increasing salt and fluid intake, or decreasing doses of hypotensive medications (Hutt-Centeno, 2019). In addition, studies in the past have used β-blockers to reduce the effects of catecholamines in the bloodstream (Chen-Scarabelli & Scarabelli, 2004). However, studies have shown that β-blockers may not be effective, and can, in some cases, worsen the symptoms of neurocardiogenic syncope (Chen Scarabelli & Scarabelli, 2004).

Midodrine is an α-adrenergic agonist, which enhances the control of the sympathetic nervous system, resulting in an increase in blood pressure by causing blood vessels to vasoconstrict, which leads to an increase in heart rate (Gurme, 2018). It works to reduce the drop in blood pressure that occurs during neurocardiogenic syncope. Yet, there has not been a largescale study of midodrine's efficacy, keeping it as a second-line option (Anstey, 2017). Therefore, this meta-analysis will quantify the magnitude of the effect of midodrine in the treatment of neurocardiogenic syncope to investigate whether the administration of midodrine actually leads to a reduction in the number of syncopal recurrences when compared to a placebo treatment. Further, this study aims to examine the reliability and functionality of midodrine.

Methods

Study Design.

Four studies were used in this meta-analysis. A meta-analysis is a statistical procedure for combining data from multiple studies. Data was entered into RevMan in order to perform the meta-analysis. The purpose of meta-analysis is to allow researchers to determine whether the effect of a treatment is consistent, which can provide an accurate depiction of the efficacy and reliability of the treatment.

Article search and identification.

Electronic databases, including PubMed, Google scholar and ScienceDirect were used to search for articles used in this study. Inclusion criteria included "neurocardiogenic syncope", "vasovagal syncope", "midodrine", "frequency of symptoms", "low blood pressure", "sympathetic stimulation". Any article that tested midodrine as a potential treatment for neurocardiogenic syncope was chosen after compiling multiple articles that fit the criteria. From the articles found, variables in each article were assessed to determine which variables were taken into consideration in the experiment. Only the articles comparing the administration of midodrine and the recurrence rate of syncope in control and test groups were included. The articles were also required to contain numerical values for the number of participants that experienced recurrence of neurocardiogenic syncope. Four of the articles matched the criteria and were included in this meta-analysis.

Data Analysis.

Two, independent, reviewers extracted the number of events when syncope recurred and the total events for both the midodrine and control group. RevMan was used to compile the data from each of the four articles, using the random effects model to calculate the pooled variance. 'Dichotomous' was selected as the data type and was analyzed using the Mantel-Haenszel statistical method. The effect of the analysis was measured using risk difference (RD). Heterogeneity was assessed using the X^2 and I^2 statistic.

Results

Publication selection.

Initially 16 publications were selected using the inclusion criteria. 10 of the 16 publications were excluded: three publications were review articles, four publications did not have a placebo comparison, one publication was a meta-analysis, and two publications used different variables. The remaining six full-length studies were then assessed for eligibility. Of the remaining publications, two publications did not provide all the required numerical data. Four studies were used in this meta-analysis (Table 1, Table 2).

Syncope recurrence.

The difference in syncope recurrence between the midodrine vs. placebo was p = 0.0004 (Figure 1). There is a reduction in recurrence of syncope when midodrine is administered where risk difference (RD) was -0.47 (95% CI -0.72, -0.21) (Figure 1). The heterogeneity present in the study was found to be $I^2 = 68\%$ and $X^2 = 9.43$ (p = 0.02) (Figure 1).

Table 1. Participants with syncope recurrence in midodrine vs. control group. Number of participants that had recurring events of syncope out of the total participants in each group for midodrine and control conditions.

Title of Study:	Midodrine		Control	
	Participants with recurrence	Total	Participants with recurrence	Total
Effectiveness of Midodrine Treatment in Patients with Recurrent Vasovagal Syncope Not Responding to Non-Pharmacological Treatment (STAND-trial)	11	23	15	23

Efficiency of Midodrine Hydrochloride in Neurocardiogenic Syncope Refractory to Standard Therapy	1	10	5	10
Usefulness of Midodrine in Patients with Severely Symptomatic Neurocardiogenic Syncope: A Randomized Control Study	6	31	26	30
The efficacy of midodrine hydrochloride in the treatment of children with vasovagal syncope	2	9	8	10

Table 2. Characteristics of the included publications— study titles, publication date, number of participants (n), the criteria each study used to assess the eligibility of participants, midodrine dose and significance.

Title of Study	Year	Number of patients (n)	Frequency of syncopal episodes prior to study	Dose of midodrine	Follow-up duration
PerezLugones et al.	2001	61	At least 1 per month	5-15 mg/ every 6h	6 months
Qingyou et al.	2006	26	At least 3 per year	1.25 mg/ 2 times daily	10 ± 8 months
Romme et al.	2011	23	At least 3 in last two years	5 mg/ 2 times daily	3 months

Sra et al.	1997	10	At least 1 per month	2.5 mg/ 3 times daily	17± 4 weeks

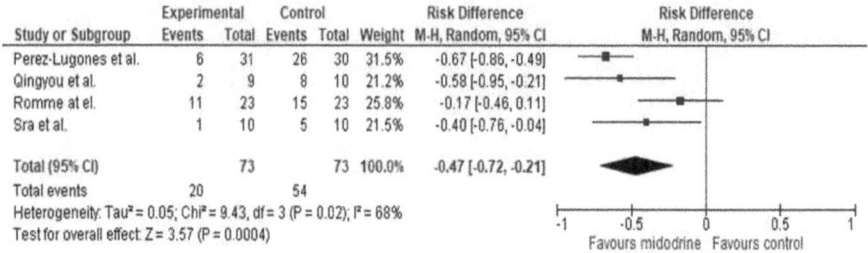

Figure 1. Risk of syncope recurrence in midodrine vs control. CI = confidence interval; IV = inverse variance method; M–H = Mantel–Haenszel method. Risk difference of less than zero favours midodrine and the risk difference of greater than zero favours control. The left panel consists of studies arranged in alphabetical order. Each study is represented by a square and the corresponding horizontal line displays the 95% confidence interval.

Discussion

This meta-analysis investigated the effect of midodrine when compared to that of a control in managing the recurrence of neurocardiogenic syncope in patients. Although some medications may be effective in managing such symptoms, another episode is likely to occur in the future. It was found that midodrine led to the decrease in both symptoms and in the frequency of syncope episodes. Qingyou et al. conducted a study which examined the effectiveness of midodrine on children with recurrent syncope (2006). They found that after the six-month follow-up period, the group given midodrine had significantly lower recurrence compared to the group given conventional therapy (Qingyou et al., 2006). Furthermore, Sra et al. also showed a significant reduction in both presyncope and syncope episodes in the group given midodrine (1997). The complete resolution of symptoms in five patients and a significant reduction of symptoms in four patients was also reported (Sra et al., 1997). In contrast, Romme et al. concluded that midodrine is ineffective in reducing the frequency of syncope episodes as no significant difference was identified between the midodrine and the placebo groups (2011). Finally, Perez-Lugones reported a significant

90

decrease in recurrence before and after midodrine was administered. It was also stated that syncope episodes that did recur in patients are likely due to the short half-life of midodrine and can be resolved by increasing the frequency of administration. This meta-analysis combined the results of these studies and showed that the group that was given midodrine had a significantly lower rate of syncope recurrence compared to the group given placebo treatment. Although the heterogeneity was significant, this is not considered a limitation to the study because there was a similar directional trend seen in all studies which favoured midodrine.

Limitations.

While the meta-analysis supports midodrine as a long-term solution for neurocardiogenic syncope it cannot compare all the variables within a study or those regarding a patient's daily life and relies only on the main effects or results, thus demonstrating a major limitation within this analysis. Furthermore, it was assumed that all the studies involved had similar participants or similar methods of obtaining participants; for example, random sampling or randomized grouping of patients into the group who attain midodrine compared to the patients who receive the placebo. Moreover, another limitation is that only one study illustrates the effects that midodrine has on children. This suggests more trials and research is needed to confirm its benefits on children suffering from neurocardiogenic syncope. However, if there is a considerable amount of dissimilarity or heterogeneity in the various studies used then a metaanalysis would not be an appropriate method to gather data, hence it will not provide the best results.

Clinical implications.

This study can have clinical implications as it supports the longterm efficacy of midodrine in the treatment of neurocardiogenic syncope. All the studies reviewed in this meta-analysis had a follow-up period of longer than three months, proving that midodrine is effective for reducing the syncopal episodes for greater than three months. Furthermore, this meta-analysis consists of studies conducted on both adults and children. This suggests that midodrine can be an effective treatment for a wide age range of patients.

Additionally, the meta-analysis can serve as a steppingstone for future investigations, such as its effects on different ethnicities, genders, diseases/disorders, etc. In particular, examining how diet may be a factor will lead to more effective and efficient medical implementations. This metaanalysis can also be utilized to examine the long-term side effects on the patients used. Specifically, if it has an impact on stress, anxiety, and overall mental health.

Conclusion

Meta-analysis of several clinical studies has shown that midodrine is an effective treatment in patients with neurocardiogenic syncope. This analysis displayed the reliability of midodrine as a treatment for patients with neurocardiogenic syncope. The clinical studies used for this study suggest that the use of midodrine compared to a placebo treatment is highly effective in reducing the recurrence of syncope. The findings of this study have the potential to make midodrine one of the first-line treatments for neurocardiogenic syncope as the metaanalysis of these clinical studies show promising results. An integrative approach to treat neurocardiogenic syncope, utilizing midodrine, can be considered as it has shown to reduce the recurrence of syncope episodes.

References

Anstey, M. H., Wibrow, B., Thevathasan, T., Roberts, B., Chhangani, K., Ng, P. Y., Levine, A., DiBiasio, A., Sarge, T., & Eikermann, M. (2017). Midodrine as adjunctive support for treatment of refractory hypotension in the intensive care unit: a multicenter, randomized, placebo controlled trial (the MIDAS trial). BMC anesthesiology, 17(1), 47. https://doi.org/10.1186/s12871-017-0339-x

Brignole M. (2007). Diagnosis and treatment of syncope. Heart (British Cardiac Society), 93(1), 130–136. https://doi.org/10.1136/hrt.2005.080713

Chen-Scarabelli, C., & Scarabelli, T. M. (2004). Neurocardiogenic syncope. BMJ (Clinical research ed.), 329(7461), 336–341. https://doi.org/10.1136/bmj.329.7461.336

Gurme M., Quan D., & Oskarsson B. E. (2020, April 21). Idiopathic Orthostatic Hypotension and other Autonomic Failure Syndromes Medication: Mineralocorticoids, Alpha adrenergic agonists, Beta-adrenergic blocking agents, Vasopressors, Erythropoietins, Gastroprokinetic agents, Anticholinesterase inhibitors, Bulk agents, Antispasmodic agents, Cholinergic agents, Phosphodiesterase inhibitors, Corticosteroids, Immune globulins. Retrieved from https://emedicine.medscape.com/article/1154266-medication#3

Higgins, J.P.T. and Thompson, S.G. (2002), Quantifying heterogeneity in a meta-analysis. Statist. Med., 21: 1539-1558. doi:10.1002/sim.1186

Hutt-Centeno, E., & Mayuga, K. A. (2018). What can I do when first-line measures are not enough for vasovagal syncope? Cleveland Clinic Journal of Medicine, 85(12), 920-922. doi:10.3949/ccjm.85a.17112

Pietrangelo, A. (2019, September 18). Presyncope: What It Is, Causes, Symptoms, and Treatment. Retrieved from https://www.healthline.com/health/presyncope

Qingyou Z, Junbao D, Chaoshu T. (2006). The efficacy of midodrine hydrochloride in the treatment of children with vasovagal syncope. J Pediatr., 149(6):777-780. https://doi.org/10.1016/j.jpeds.2006.07.031

Richard N. Fogoros, M. (2020, March 16). Vasovagal Syncope: A Common Cause of Fainting. Retrieved from https://www.verywellhealth.com/vasovagal-cardioneurogenic-syncope1746389

Romme, J. J. C. M., Dijk, N. V., Go-Schon, I. K., Reitsma, J. B., & Wieling, W. (2011). Effectiveness of Midodrine treatment in patients with recurrent vasovagal syncope not responding to non-pharmacological treatment (STAND-trial). Europace, 13(11), 1639– 1647. doi: 10.1093/europace/eur200

Ward, C. R., Gray, J. C., Gilroy, J. J., & Kenny, R. A. (1998). Midodrine: a role in the management of neurocardiogenic syncope. Heart, 79(1), 45–49. doi: 10.1136/hrt.79.1.45

Should internationally trained doctors be able to practise during the covid crisis?

Austin Mardon and Gina Schopfer

As Covid-19 cases continue to rise in Canada, the question of whether our medical systems will reach a point of inability to respond to all cases continues to loom in the back of our minds. Considering Italy: the pandemic's epicenter throughout March, it is a realistic possibility that Canadians must be aware of in order to properly apply preventative measures to our day-to-day lives, and think about what we can use to tackle the spread as quickly and effectively as possible.

As of mid-March, Ontario has begun allowing internationally trained medical graduates who have passed their exams to practise in Canada to apply for a 30-day medical license to help fight growing virus cases. This Supervised Short Duration Certificate allows these professionals to practise under supervision at public medical centres.

The criteria for acceptance include:

- Graduation from a medical school in Canada, the United States, or a school listed in the World Directory of Medical Schools.
- Having practiced medicine, graduated medical school, or passed Medical Council of Canada exams within the past two years.
- Having secured a spot working in a medical centre.
- Having found a licensed physician to act as a supervisor.

Alberta is slower to follow Ontario's lead, but the potential to do so is there. Allowing these professionals to practise their skills while providing them with an income and purpose during these uncertain times offers too many benefits to simply sweep the idea under the rug. Considering the lengthy and bureaucratic process medical professionals undergo to practice in Alberta today, it would be well worth looking into making the process more efficient, or implementing a Supervised

94

Short Duration Certificate as Ontario has done during a time of true necessity. This is especially true in Southern Alberta and the Calgary area, and cases of Covid-19 go up by the hundreds day by day.

Postmedia has reported that there are hundreds of internationally trained doctors in the province able to help. Rather than undergoing multiple credential assessments, knowledge exams, and practical assessments, it would be beneficial to allow these professionals to contribute to providing attention and care to the rising cases before Alberta hits its peak.

There is a question of whether our current economy could accommodate this change, and there is still a lot of red tape to cut through. Nonetheless, it does no harm to consider our options and prioritize the health of Canadians first and foremost.

***

Gina Schopfer is a graduate of MacEwan University's Bachelor of Communication Studies program. She is a researcher and writer for the Antarctic Institute of Canada. Austin Mardon is an assistant adjunct professor at the John Dossetor Health Ethics Centre at the University of Alberta, as well as an author, community leader, and advocate for the disabled. He founded the Antarctic Institute of Canada and has been awarded the Order of Canada.

Links:

https://www.cbc.ca/news/canada/toronto/internationally-trained-doctors-covid-19-1.5519881

https://calgarysun.com/opinion/leong-alberta-must-put-foreign-trained-doctors-on-fast-track

How does wearing a mask protect you and others around you?

Andrew Clement[1], Jasrita Singh[1,2], Austin A. Mardon[1,3]

Affiliations:

1. Antarctic Institute of Canada (AIC), Alberta, Canada.
2. Faculty of Health Sciences, McMaster University, Hamilton, Ontario, Canada
3. John Dossetor Health Ethics Centre, University of Alberta, Edmonton, Alberta, Canada

In December 2019, a novel zoonotic virus began to spread from Wuhan to other areas of China and rapidly made its way across the globe. According to the John Hopkins Coronavirus Research Center, the severe acute respiratory syndrome coronavirus 2 (SARS-CoV-2) is responsible for infecting over 18 million people around the world as of August 5th, 2020. COVID19 is highly contagious, with a reproductive number (R0) between 2-3, with some studies reporting that it is as high as 6.49[1]. The reproductive number illustrates how infection an agent is, i.e., each COVID-19 positive individual will potentially infect 2 or 3 other people.

COVID-19 causes a large variety of symptoms depending on the severity of the infection. However, the most well-known symptoms arise from the virus targeting the lungs[2], kidneys[3], and cardiovascular system. We know that symptomatic patients can transmit the virus; however, recent studies have indicated that pre-symptomatic and asymptomatic individuals may be transmitting COVID-19[4]. Pre-symptomatic individuals are infected and are spreading the virus before developing symptoms, where asymptomatic individuals are infected but do not present symptoms throughout their infection. An estimated 15% of infected people are asymptomatic[5], creating a problem in controlling the spread of the disease. Some measures include physical distancing, proper hand hygiene, and mask-wearing. These measures have shown to decrease the spread of the virus significantly[6].

Currently, the modes of transmission for COVID-19 are not clear. When the virus began to circle the globe, the thought was that the main modes of transmission were through inanimate objects (fomites) that symptomatic people interacted with and through liquid droplets[7]. New evidence suggests the possibility of airborne transmission through aerosols[8,9]. Some studies conducted in COVID-19 wards discovered airborne viral genetic material; however, they did not prove capable of infection[10,11]. In direct contrast, other studies discovered the possibility of a viable virus remaining suspended in the air for up to 16 hours under experimental conditions[12]. Evidently, given the novel nature of the virus and its effects, COVID-19 research is continually evolving. Hence, is it essential to take necessary precautions to prevent its spread.

As mentioned, many interventions to decrease the spread have been taken, one of which is wearing a mask. According to the WHO, two different types of masks are readily available to the public – the non-medical cloth masks and disposable medical masks. The disposable masks are made of 3-4 layers of fibers that can filter particles as small as 0.1μm, preventing larger particles' transmission. These sorts of masks are standardized and tested to ensure effectiveness. In contrast, the non-medical cloth masks do not need to be tested; according to the government of Canada, these masks must be a minimum of two tightly woven fabrics (such as cotton or linen) and can be washed and reused[13]. Masks have since been commonplace all over the world. Therefore, it is essential to practice proper handling and care while putting on, wearing, removing, and cleaning masks, as improper handling can increase the risk of contamination and infection[14,15]. Some common errors include, but are not limited to:

- Touching the mask or face after putting it on
- Wearing a loose mask
- Reusing a single-use mask
- Removing the mask to communicate
- Pulling the mask down to your neck
- Uncovering your nose
- Allowing someone else to handle your mask
- Using a mask that is damaged or dirty

The WHO has come out with videos on how to do this with medical masks and non-medical masks[16]. The steps to put on and remove a mask are as follows
:

1. Wash your hands for 20 seconds with soap or hand sanitizer
2. By grabbing the ear loops, inspect the mask for damage or if dirty
3. Place the mask so that your mouth and nose are covered and secure the loops behind your ears
4. Adjust the mask so that your mouth and nose are covered and that that there are no gaps on the sides
5. If available, pinch the metal strip to mold it to your nose
6. To remove the mask, wash your hands for 20 seconds with soap or hand sanitizer
7. Take the side ear loops and pull the mask off away from your face
8. Place the mask in a clean bag, or toss in the garbage for non-reusable masks
9. Wash your hands for 20 seconds with soap or hand sanitizer

Another problem that arises with non-medical cloth masks is cleaning. Masks should be cleaned daily by removing them from their bag by the ear loops and washed with detergent in a hot cycle or by hand in hot water. Masks should then be hung out to dry completely before the next use. Masks can provide a sense of community where people come together to protect each other. Masks can also serve as a mode of expression; customizable non-medical masks are becoming more popular and may manifest culture and self.

COVID-19 has shown that it is capable of rapid transmission, even when an infected individual presents with no symptoms[5]. This presents many problems as people who do not know they are sick can infect others; therefore, it is imperative to follow measures to prevent the spread, such as wearing a mask even if you don't present with symptoms.

References:

1.   Liu, Y., Gayle, A., Wilder-Smith, A. & Rocklov, J. The reproductive number of COVID19 is higher compared to SARS coronavirus. J. Travel Med. 27, (2020).

2.   Yuki, K., Fujiogi, M. & Koutsogiannaki, S. COVID-19 pathophysiology: A review. Clinical Immunology 215, 108427 (2020).

3.	Martinez-Rojas, M. A., Vega-Vega, O. & Bobadilla, N. A. Is the kidney a target of SARSCoV-2? American journal of physiology. Renal physiology 318, F1454–F1462 (2020).

4.	Rothe, C. et al. transmission of 2019-NCOV infection from an asymptomatic contact in Germany. New England Journal of Medicine 382, 970–971 (2020).

5.	Byambasuren, O. et al. Estimating the extent of asymptomatic COVID-19 and its potential for community transmission: systematic review and meta-analysis. medRxiv (2020).

6.	Chu, D. K. et al. Physical distancing, face masks, and eye protection to prevent person-toperson transmission of SARS-CoV-2 and COVID-19: a systematic review and metaanalysis. Lancet 395, 1973–1987 (2020).

7.	Ong, S. W. X. et al. Air, Surface Environmental, and Personal Protective Equipment Contamination by Severe Acute Respiratory Syndrome Coronavirus 2 (SARS-CoV-2) from a Symptomatic Patient. JAMA - Journal of the American Medical Association 323, 1610– 1612 (2020).

8.	Fears, A. C. et al. Persistence of Severe Acute Respiratory Syndrome Coronavirus 2 in Aerosol Suspensions. Emerg. Infect. Dis. 26, (2020).

9.	Cai, J. et al. Indirect virus transmission in cluster of COVID-19 cases, Wenzhou, China, 2020. Emerg. Infect. Dis. 26, 1343–1345 (2020).

10.	Guo, Z. D. et al. Aerosol and Surface Distribution of Severe Acute Respiratory Syndrome Coronavirus 2 in Hospital Wards, Wuhan, China, 2020. Emerg. Infect. Dis. 26, 1586–1591 (2020).

11.	Chia, P. Y. et al. Detection of air and surface contamination by SARS-CoV-2 in hospital rooms of infected patients. Nat. Commun. 11, 1–7 (2020).

12.	Van Doremalen, N. et al. Aerosol and surface stability of SARS-CoV-2 as compared with SARS-CoV-1. New England Journal of Medicine 382, 1564–1567 (2020).

13. COVID-19: Non-medical masks and face coverings - Canada.ca. Available at: https://www.canada.ca/en/public-health/services/diseases/2019-novel-coronavirusinfection/prevention-risks/about-non-medical-masks-face-coverings.html. (Accessed: August 4th 2020)

14. Kwon, J. H. et al. Assessment of Healthcare Worker Protocol Deviations and SelfContamination during Personal Protective Equipment Donning and Doffing. Infect. Control Hosp. Epidemiol. 38, 1077–1083 (2017).

15. Zamora, J. E., Murdoch, J., Simchison, B. & Day, A. G. Contamination: A comparison of 2 personal protective systems. CMAJ 175, 249–254 (2006).

16. When and how to use masks. Available at: https://www.who.int/emergencies/diseases/novel-coronavirus-2019/advice-forpublic/when-and-how-to-use-masks. (Accessed: August 4th 2020)

Burdens and Tradeoffs: Options for Online Examinations during COVID-19

Mehvish Masood, Austin Mardon

In tradeoffs between privacy, comfort and equitable examinations, online schooling caused by COVID 19 has raised concerns about effective testing conditions in university settings. Professors grapple with having to choose between online proctoring software, using Zoom calls or having no proctoring at all for evaluations. A lack of consensus has emerged with which is regarded to be the most effective and many students have had to adapt to a variety of these testing conditions.

Online proctoring software, predominantly Proctortrack, have been used by students. By having access to a student's video camera, microphone and computer programs, this software can guarantee academic integrity by ensuring students are writing the test and are not cheating by accessing notes or searching for answers online.

While proctoring software maintains academic integrity, safety and privacy concerns have been raised with this software. Data from students of Western University and Queen's University have both been subject to a security breach (Harmsworth; Lupton). While Proctortrack and university officials have insisted that security is of utmost importance, students still fear the impact this software will have on their privacy (Lupton). This issue has incited a petition to be with the concerns raised at Western University, which currently has over 10,000 signatures (Iyayi).

Furthermore, issues have been raised about the accessibility proctoring software give to students, especially neurodivergent individuals. For example, Proctortrack works by "tagging" individuals that either has unusual behaviour such as eye movements, body movements, moving out of the frame etc. The professor then checks these "tagged" moments to ensure cheating has not occurred. While this may seem insignificant, it creates a bias towards neurodivergent individuals, who sometimes require movement to function under testing conditions, to be tagged disproportionately. This disrupts the accuracy of the software and raises certain questions on the equitable nature.

Washroom breaks have been of concern with this software. While professors have the option of not "tagging" individuals that leave for the washroom on this software, many choose not to in exams as a means to maintain academic integrity. This forces many students to not have the option to have washroom breaks for up to 3 or 4 hours at a time. Not only can this impair a student's ability to do examinations, it disproportionately harms those that are subject to health issues that require them to have washroom breaks.

An alternative option that has been proposed for proctoring is Zoom calls. Professors monitor students for the duration of the examination or test to ensure that the individual is the one that is writing the test. Notably, academic integrity is maintained at a lower level on Zoom calls compared to proctoring software because professors are unable to access what a student is searching up on their devices, which gives the students free rein to check anything on the internet. A major disadvantage of Zoom calls is that examinations in this setting are only able to be conducted with smaller class sizes. The inability for professors to track many students at a time, along with the participation limit of 100 on Zoom calls, makes it such that up to 50 students can be proctored at the same time. This makes it difficult for classes that have large class sizes, which can go higher than 500 to 1000 students, to use Zoom calls as a viable option for exams.

The last option professors have is not doing proctoring at all. While this is not preferable due to the inability to check academic integrity, it allows the bypassing of the concerns of the other methods. Notably, the ability to cheat through using notes and the internet is not the only concern for professors for this method. Instead, professors fear that another person that is not the student is writing the exam. Professors have responded to these fears by making students accept honour pledges to not cheat with the legitimate threat of severe academic penalties and expulsion if they are found not to follow it. However, even with such pledges, the inability to check if students follow through with not cheating prevents professors from wanting to choose this method.

In the circumstances that professors allow for non-proctored exams, professors have had several methods to maintain academic integrity. Firstly, professors have opted to either make exams more difficult and highly application based ensuring that even when notes and the internet can be accessed, students that have not studied the content will be unable to be successful. Secondly, professors have opted to do "linear" exams, in which students are unable to go back to previous questions after submitting answers. This ensures that students do not have abundant time to cheat by contacting others or search through the internet. Those that understand the content would be able to be successful with linear examinations. The last option that professors have pursued is putting less "weight" on the exam and instead opting for other smaller assignments or quizzes during the semester itself. For example, instead of having an exam worth 40% of a student's grade, a student may have an exam worth 25% and 5 quizzes worth 3% each. While many professors have pursued shifting the weight option, it has created a strain on students that have to now balance smaller quizzes during the semester, which is burdensome especially during a time where students are struggling to adjust to online learning.

During this shift to online education during COVID-19, many students have had to adapt to the changing environment with which education is taught. The online adjustment of exams has different methods of proctoring, each with potential advantages and disadvantages. These need to be considered when choosing how to deal with examinations to ensure the best practices to maintain high standards of education, while dealing with online learning during COVID-19.

Mehvish Masood is a student working towards a Medical Science degree at Western University. Austin Albert Mardon is an Order of Canada member and Fellow of the Royal Society of Canada. Catherine Mardon is a retired attorney and advocate for the disabled.

References

Harmsworth, Julia. "Proctortrack Suspends Service Following Breach, Impacting FEAS & Smith." The Journal, https://www.queensjournal.ca/story/2020-10-22/news/proctortracksuspends-service-following-breach-impacting-feas-and-smith/. Accessed 22 Jan. 2021.

Iyayi, Mudia. "2020 in Review: The Good, the Bad and Everything We'd like to Forget." The Gazette , https://www.facebook.com/westerngazette, 18 Jan. 2021, https://westerngazette.ca/culture/2020-in-review-the-good-the-bad-and-everything-wed-like-toforget/article_997f41c4-503a-11eb-ac5a-13b3e44461a5.html.

Lupton, Andrew. "Western Students Alerted about Security Breach at Exam Monitor Proctortrack." CBC, 15 Oct. 2020, https://www.cbc.ca/news/canada/london/western-studentsalerted-about-security-breach-at-exam-monitor-proctortrack-1.5764354.

The Link between COVID-19 and Cardiovascular Diseases

Sriraam Sivachandran, Navneet Kang, Jasrita Singh,
Austin Mardon

Abstract

COVID-19 was deemed a pandemic in March of 2020 and since then
the number of cases and deaths worldwide has greatly increased. It has
been researched on how COVID-19 affects the various bodily systems
and one such system is the cardiovascular system. It has been seen
that people with certain cardiovascular diseases are at higher risk of
getting infected by the virus. Many of the older citizens that were
infected with the virus in China did in fact suffer from hypertension.
Even though, it has only been a year since the virus has affected us
globally, it is important to further investigate how the virus affects the
cardiovascular system and other bodily systems.

The onset of the SARS-CoV-2 pandemic revolutionized the world at
breakneck speed. The rapid increase in the cases along with the lack of
knowledge on the virus led to a continuous uphill battle in 2020. When
a novel disease comes about, it is important to analyze the statistics that
pertain to different countries as this will allow for the adoption of different
countries' strategies. As of January 15, 2021, the government of Canada
states that there have been 688, 891 cases with 593, 397 individuals
recovered and 17, 538 deaths. It is also important to note that majority
of the cases and deaths were reported in Ontario and Quebec[1]. Various
scientific outlets such as the World Health Organisation (WHO) and
Centers for Disease Control and Prevention (CDC) have provided the
world with important knowledge surrounding the virus. The WHO
and CDC have provided the general public with evident symptoms
of COVID-19, including but not limited to fever, dry cough, fatigue.
However, it is important to understand the impact of COVID-19 on
the cardiovascular system. Clerkin et al stated that, "Among patients
with COVID-19, there is a high prevalence of cardiovascular disease"[2].

Extensive research was applied in order to better understand the effects of COVID-19 on cardiovascular diseases, such as hypertension.

Firstly, it is important to understand the organs that comprise the cardiovascular system along with the numerous pathways that are involved in carrying out its function. The three major components of the cardiovascular systems are the heart, the blood vessels, and the blood. The heart pumps blood around the body, which is crucial as the blood contains important nutrients and oxygen. The pumped blood is then carried to different parts of the body using blood vessels. Blood that is being transported away from the heart uses arteries and blood that is being transported to the heart uses veins. Simply, the cardiovascular system can be understood as a bodily system that aims to maintain homeostasis.

There have been research papers that explained the connection between COVID-19 and the cardiovascular system. Hypertension, or high blood pressure, is a cardiovascular disease that has been linked to the virus. In light of COVID-19, there have been several questions on its effect on people with hypertension and whether it can lead to hypertension. Firstly, hypertension is prevalent in older populations and was reported that many of the older patients infected with COVID-19 in China and Italy did in fact suffer from hypertension[3]. However, this may seem like confounding evidence as many older citizens already suffer from hypertension.

Hypertension as a symptom of the virus was also seen more predominantly in more serious COVID-19 cases in Wuhan, China. Due to the fact that the older population is primarily affected, there is an increased prevalence of hypertension. While hypertension may be an indicator of COVID-19, clinical trials must be done to confirm this notion.

1

(Government of Canada, 2020)

2

(Clerkin et al., 2020)

3

(Shibata et al., 2020)

Given that COVID-19 is such a novel situation, it is not a surprise that scientists have not reached definitive conclusions regarding cardiovascular diseases and their link to COVID-19.

These links could be the onset of certain cardiovascular diseases or the increased infection of COVID-19 due to patients pre-existing conditions. In the future, it is important to fully understand how COVID-19 can impact the cardiovascular system as this may make it simpler for healthcare officials to identify certain markers in their patients.

References

Canada, P. H. A. of. (2020, April 19). Epidemiological summary of COVID-19 cases in Canada [Datasets;statistics;education and awareness]. Aem. https://health-infobase.canada.ca/covid-19/epidemiological-summary-covid-19-cases.html

CDC. (2020, May 19). High Blood Pressure Symptoms, Causes, and Problems | cdc.gov. Centers for Disease Control and Prevention. https://www.cdc.gov/bloodpressure/about.htm

Clerkin, K. J., Fried, J. A., Raikhelkar, J., Sayer, G., Griffin, J. M., Masoumi, A., Jain, S. S., Burkhoff, D., Kumaraiah, D., Rabbani, L., Schwartz, A., & Uriel, N. (2020). COVID-19 and Cardiovascular Disease. Circulation, 141(20), 1648–1655. https://doi.org/10.1161/CIRCULATIONAHA.120.046941

Shibata, S., Arima, H., Asayama, K., Hoshide, S., Ichihara, A., Ishimitsu, T., Kario, K., Kishi, T., Mogi, M., Nishiyama, A., Ohishi, M., Ohkubo, T., Tamura, K., Tanaka, M., Yamamoto, E., Yamamoto, K., & Itoh, H. (2020). Hypertension and related diseases in the era of COVID-19: A report from the Japanese Society of Hypertension Task Force on COVID-19. Hypertension Research, 1–19. https://doi.org/10.1038/s41440-020-0515-0

Siripanthong, B., Nazarian, S., Muser, D., Deo, R., Santangeli, P., Khanji, M. Y., Cooper, L. T., & Chahal, C. A. A. (2020). Recognizing COVID-19–related myocarditis: The possible pathophysiology and proposed guideline for diagnosis and management. Heart Rhythm, 17(9), 1463–1471. https://doi.org/10.1016/j.hrthm.2020.05.001

Author Biographies:

Austin Albert Mardon, CM Ph.D. is an author, community leader, and advocate for mental health. He is an assistant adjunct professor at the John Dossetor Health Ethics Centre at the University of Alberta. In the mid 80's, he founded and today still directs the Antarctic Institute of Canada, a non-profit entity based in Edmonton, Alberta. He is also an Order of Canada member and Fellow of the Royal Society of Canada.

The Pet Transience During the COVID-19 Pandemic

Leah Sarah Peer, Lina Lombo, Jasrita Singh,
Daivat Bhavsar, Austin Mardon

Sharing our lives with pets not only makes us healthier in normal times but also allows them to serve as comfort companions during stressful times such as the Coronavirus pandemic. This non-human companion provides emotional support and endless love as well as decreases psychological arousal and stress, by eliciting physiological changes that make us feel better.[1] These include increased dopamine and oxytocin levels in humans which are the "happy hormones" that cause us to experience positive emotions.[9] Additionally, lowered levels of epinephrine, norepinephrine play a role in the regulation of the sympathetic nervous system that is responsible for the body's "fight or flight" responses, as well as reduced average arterial and systolic blood pressures.[7]

With global lockdowns and an international health crisis, pandemics naturally bring stress, fear, and anxiety into people's lives. As such, there were a variety of differential responses of people globally from scurrying to purchasing toilet paper in the USA to abandoning pets in Wuhan; however, in Canada, quite the opposite occurred[3]. There was a surge in pet adoptions. Many animals from shelters had a place to call home, and consequently, the mental health of those who stepped up to care for these animals improved. In this way, companion animals may provide effective stress mitigating strategies that play a role in maintaining healthy protective behaviours crucial to keeping strong during the pandemic. Furthermore, one of the first studies supporting the health benefits of pets by Friedmann et al, in 1980 indicated that one year after heart patients were discharged from a coronary care unit, pet owners were more likely to be alive than nonowners.[11] By reducing stress and improving overall mental health, companion animals contributed to their owners survival and as such play an important role in human development. For this reason, health professionals have incorporated animals into therapeutic work with ADHD children or with patients at mental institutions, simply because of the ameliorative nature of these interactions.[11]

Additionally, with the separation of families halfway across the globe, pets were often the only companion many resorted to as a source of comfort to loneliness. As a result, companion animal breeders saw an increasing breeding demand for domestic animals. Without these pets, millions would have been isolated. The pets' presence as physical beings within reach made the absence of human touch a little less onerous. For instance, Dr. O'Dair, a family physician, always wanted to adopt a pet but never found the time.[4] As she began visiting patients via telehealth platforms at home, her desire to adopt a dog seemed more plausible. Following through with her goal, she adds "I don't know what I would do without the company of my dog, she has kept me going".[4] With the companion animal, O'Dair felt capable navigating the radical lifestyle changes caused by the pandemic.

Like O'Dair, the disruption of routines led to a new way of living and a pet's presence gave purpose; purpose to care for an animal, be it feeding, walking, or otherwise playing with them on a walk outside. Evidently, pet ownership requires a long-term commitment as interactions with domestic animals naturally strengthen over time.

Not everyone is able to afford the adoption of animals and although it is a great investment, it is a hefty one.[4] With many laid off, caring for themselves alongside a pet is more challenging[3]. Physicians and professionals who are able to work from home remain unaffected financially. However individuals living off their emergency savings, on the contrary, are counting on every penny towards their survival and sustenance that caring for an animal is out of budget.

As a solution to the cost associated with raising an animal, scientists have discussed the role of artificial pets and if they can ever replace real ones[5]. In Florida, The Department of Elder Affairs provided artificial pets, therapeutic robots acting as replicas of real furry animals, to senior adults. The motivation behind the experiment was for these robots to act as companion animals who would aid the elderly dealing with their emotions dealing with the end of life[5]. Therapeutic robotic pets have also solved the more tedious duties involved with caring for an animal and its hygiene but are these robots any different to a toy?

Besides a pet's innate companionship, it's the emotional connection that humans seek. From a lifeless robot enacting as a domestic animal, the joy of a real animal is inexplicable in comparison. Love that comes with pets is not the same as an artificial creation of love, making the differences strikingly evident. Therefore, pets encourage people to be healthy, ward off the loneliness and mental health struggles exacerbated by the pandemic. At times, a dog or cat's presence is the only difference between an isolated person and despair.

Non-human companions complement our well-being and therefore, when developing programs supporting isolated citizens, or hospitalized children, we need to remember the value these animals bring[1]. Moving forward, the government needs to consider food security not just for humans but also for their non-human counterparts to prevent the possibility of a tsunami of pet abandonment due to an inability to afford food or care as was the case in Wuhan. Volunteer Li Heng of the Furry Angels Haven opened arms to 67 dogs and 40 cats in her small apartment when abandonment skyrocketed. Acting out of her instinct and concern, she did her best to preserve the safety and well-being of these animals[8].

The adoption of chronically ill animals during the pandemic has been heartwarming. Several with more free time on their hands and sufficient financial resources have stepped up to care and serve animals in need, especially of the vulnerable, the sick or older pets that may have not found a home otherwise. The response by community members to support local shelters has been enormous in some cases, as many as 400 applications per dog.[12] In Toronto, Redemption Paws, a rescue group saw 600 dogs adopted or fostered in 2020.[12] The little compassion and care ultimately reminds us of our humanity, and life's ephemerality.

With life shifting post-covid-19, animal caretakers will have to assess their changes. Fortunately, reputable shelters have weighed in applicants' schedules post-covid to ensure animals are provided with stability, and long term support. Many match their adopters' lifestyles to stay with their owners as they return to a post-pandemic schedule.

Another problem on the rise is increased aggression from pets due to prolonged lockdowns and unusually long hours with owners.[6] Dogs for example are biting, barking and owners are on the verge of giving them up again.[6] Others state that dogs have become anxious about people out on walks and thus are showing fearful behaviour to humans who are not their owner. Moreover, many puppies adopted were not trained to deal with strangers and are still not accustomed to people[6]. With more than one dog in households, these animals will act aggressively to each other, fighting for their owners' attention. With children wanting to play with puppies they will lack proper sleep or rest and this will cause them to be uneasy and difficult to deal with. The new staple attire of face masks also have older dogs wary as the lack of facial expressions visible to these species have them doubtful of others.[6] Another concern is the separation anxiety many pets will face once owners head back to work.

While adopting and caring for pets during a pandemic is challenging, if given considerable thought, support, it is very rewarding both for the animal and humans in contact.

References

1. Chia-Chun Tsai, Erika Friedmann & Sue A. Thomas (2010) The Effect of AnimalAssisted Therapy on Stress Responses in Hospitalized Children, Anthrozoös, 23:3, 245-258, DOI: 10.2752/175303710X12750 451258977

2. Headey, B., Grabka, M.M. Pets and Human Health in Germany and Australia: National Longitudinal Results. Soc Indic Res 80, 297–311 (2007). https://doi.org/10.1007/s11205-005-5072-z

3. Qiao Huang, MPH, Xiang Zhan, BE, Xian-Tao Zeng, MD, MPH, PhD, COVID-19 pandemic: stop panic abandonment of household pets, Journal of Travel Medicine, Volume 27, Issue 3, April 2020, taaa046, https://doi.org/10.1093/jtm/taaa046

4. 4. L.F. Carver Assistant Professor & Privacy and Ethics Officer at the Centre for Advanced Computing. "How the Coronavirus Pet Adoption Boom Is Reducing Stress." The Conversation, 2 Dec. 2020, www.theconversation.com/how-the-coronavirus-petadoption-boom-is-reducing-stresa.s-138074

5. Kim, Allen. "Some Florida Seniors Isolated with Alzheimer's and Dementia Due to the Pandemic Are Getting Robotic Therapy Pets." CNN, Cable News Network, 27 Apr. 2020, www.cnn.com/2020/04/27/us/therapy-robot-pets-wellness-trnd/index.html.

6. Storrar, Krissy. "Charities Fear Pets Are Becoming Stressed during Lockdown Due to Spending Too Much Time with Owners." The Sunday Post, 3 Nov. 2020, www.sundaypost.com/fp/charities-fear-pets-are-becoming-stressed-during-lockdowndue-to-spending-too-much-time-with-owners/.

7. Beetz, Andrea et al. "Psychosocial and psychophysiological effects of human-animal interactions: the pos-

sible role of oxytocin." Frontiers in psychology vol. 3 234. 9 Jul. 2012, doi:10.3389/fpsyg.2012.00234

8. Campbell, Charlie. "Chinese Shelter Seeks to Rehome Pets Abandoned in Pandemic." Time, Time, 8 Dec. 2020, time.com/5916962/animal-shelter-wuhan-china-petscoronavirus/.

9. Healthline Media, 30 Aug. 2018, www.healthline.com/health/love-hormone .

10. "Epinephrine vs. Norepinephrine: Differences, Functions, and High Levels." Medical News Today, MediLexicon International, www.medicalnewstoday.com/articles/325485

11. O'haire, Marguerite. "Companion Animals and Human Health: Benefits, Challenges, and the Road Ahead." Journal of Veterinary Behavior, vol. 5, no. 5, 2010, pp. 226–234., doi:10.1016/j.jveb.2010.02.002.

12. Anderssen, Erin. "Year of the Dog: In 2020, Furry Friends Were Just What We Needed to Make It through the Pandemic." The Globe and Mail, 19 Dec. 2020, www.theglobeandmail.com/life/article-year-of-the-dog-in-2020-furry-friends-werejust-what-we-needed-to-make/.

Leah Sarah Peer is an independent Article Writer associated with the Antarctic Institute of Canada. She is a medical student in Montreal, Quebec and a strong advocate for human rights and global health.

Lina Lombo is an undergraduate student at the University of Western Ontario with a background in the medical sciences.

Jasrita Singh is the Program Director for the Antarctic Institute of Canada's writing program, Sharpen the Quill, and is an undergraduate student in the Faculty of Health Sciences at McMaster University.

Daivat Bhavsar is the Senior Team Lead for the Antarctic Institute of Canada's writing program, Sharpen the Quill, and is an undergraduate student in the Faculty of Health Sciences at McMaster University.

Austin Albert Mardon is an Order of Canada member, Fellow of the Royal Society of Canada, and the director of the Antarctic Institute of Canada charity.

Victory Gardens: a thing of the past, or a facet of our future?

Austin Mardon and Gina Schopfer

This week, agriculture minister Marie-Claude Bibeau encouraged Canadians to consider growing war-era victory gardens: produce grown on private residences to reduce pressure on public food supply and supplement **provisions** that may be harder to access in the coming months. The victory garden initiative follows outbreaks at the High River's Cargill meat processing plant, resulting in over 300 new COVID-19 cases in southern Alberta.

"I don't worry though that we won't have enough food to eat in Canada. We might have some challenges in terms of variety and maybe on prices at this certain point, but it's always good to learn how to grow food."

The opportunity presents itself now as the snow melts and we approach our second month of spring. WWII era came with pamphlets that explained gardening basics and information on what to plant. Today, with information constantly coming up on our devices, finding useful tips can at times feel a little overwhelming. However, cultivating a victory garden offers benefits like lowering trips to the grocery store, and providing us with a hobby and opportunity to get out of the house and get some fresh air, so it's definitely worth a shot

Since we aren't all horticulture experts, here are some tips to produce success (and delicious home-grown fruits and vegetables):

1. Be selective

Before you can plant, you must choose what you'd like to grow. Research and ask yourself: will this grow well in my climate? Will I get use out of this? How difficult is this to maintain? My friend's Grandmother says, "the easiest things to grow in a garden are the things you're excited to eat," so try to grow things you can look forward to! If this is your first garden, there's nothing wrong with keeping things simple and sticking to a few different types of seeds.

Look for gardening centres offering curbside pickup or delivery when purchasing your seeds.

2. Start seeds indoors

Most annual vegetables should be sown indoors about six weeks before the last frost in the area. Pay attention to weather forecasts. As we are approaching May, now is likely the time to start.

To start, pot with a seed-starting mix. Ensure the containers have drainage holes, and plant seeds at a depth that is two times the width of the seed. Set containers in a warm area, and make sure to keep the mix moist. Once seedlings emerge, place the containers in a sunny space.

3. Use space efficiently

Separate fast-growing and slow-growing crops. Be aware of each plant's likelihood to spread and how far. If you have limited space, you will only be able to choose a select few options to grow. Note the space you have available and strategically plan how to fill the space.

4. Plant flowers

Flowers attract bees and other pollinators, which will significantly improve the likelihood of a full garden. As an added bonus, flowers can add beautiful pops of colour to your outdoor space and create a great atmosphere.

5. Fertilize your soil

Fertilizer is very important because it contains plant nutrients that help plants grow. It can be bought at any business with a gardening centre, so finding options for curbside pickup shouldn't be a problem. Making your own compost is also a great way to fertilize your plants while helping reduce household waste and saving you some money.

A final piece of gardening advice is to take notes on what you've grown each year, what worked, what didn't, and what you may want to consider trying next year. If you still don't feel fully confident, keep in mind that any questions you may have will have answers that you can find online. The technological advancements we have today serve us as a readily available outlet for information and contact with the outside world. Daniel and Alexandria Randell live in Calgary and have been using their website and blog "My Vintage Lifestyle" to provide entries on how to manage when things are scarce, and wartime recipes for lean times. Using our new "free" time productively. Though we have to find ways to adapt to our new circumstances, we are in this together and all play a part in how we do so.

***

Gina Schopfer is a graduate of MacEwan University's Bachelor of Communication Studies program. She is a researcher and writer for the Antarctic Institute of Canada. Austin Mardon is an assistant adjunct professor at the John Dossetor Health Ethics Centre at the University of Alberta, as well as an author, community leader, and advocate for the disabled. He founded the Antarctic Institute of Canada and has been awarded the Order of Canada.

Links:

https://www.cbc.ca/news/politics/food-security-pandemic-freeland-1.5534372

https://calgaryherald.com/news/province-to-update-covid-19-numbers-at-330-p-m/

https://www.tourismsaskatoon.com/blog/post/saskatoon-gardening-experts-sharetips/?utm_source=facebook&utm_medium=social&utm_campaign=StayConnected&utm_

Menstruation Through the Eyes of the Homeless and Trans People

Austin Mardon and Gina Schopfer

One of the greatest underlooked issues of today's generation is the issue of menstrual hygiene. Most people- men or women- will shy away from or downright ignore the topic of menstruation when it is brought up. They will deny it exists, they will deem the conversation inappropriate and they will shun away anyone who wants to talk about it. Yet this phenomenon affects most if not all members of the fairer sex. The fact that we cannot even openly discuss the process of menstruation displays the limitations faced when considering menstrual hygiene. When a woman is on her cycle, she can go to the store and pick from numerous different styles, brands and shapes of menstrual hygiene products. However, what we forget is that it is not as simple for those women that are facing homelessness or are trans. Moreover, with the pandemic raging around the world today, this basic human right seems more like a luxury for the more fortunate.

For this reason, a student group and affiliate group at Ryerson University named Women in Information Technology Management, or WITM, have made it their mission to understand the needs of women facing homelessness and trans people. Moreover, they strive to supply them with the products needed to remain safe and hygienic. Unhygienic menstrual products can lead to a plethora of maladies and risks including but not limited to fungal infections, urinary infections, rashes and more. Many scientific journals, including "Are Unhygienic Practices During the Menstrual, Partum and Postpartum Periods Risk Factors for Secondary Infertility?" by Tazeen Saeed Ali, Neelofar Sami and Ali Khan Khuwaja, highlight the major risks associated with unhygienic menstruation practises. These include rewashing and reusing specific cloths to act as pads as well as not properly cleansing one's body during their menstrual cycle. These practises can lead to major issues such as secondary infertility, caused by drastic infections of the female reproductive organs. Unfortunately for homeless women, sometimes these unhygienic practises are their only option, as they do not have clean products available at their disposal. Moreover, trans individuals are often faced with discrimination when trying to utilize their right to go to the washroom of their gender. Thus, both parties are at higher risk for infections and other health risks because they cannot access the correct products as readily as the more privileged female populations. The inability to speak on this matter and raise proper resources results in these parties continually suffering every month. For homeless women, on top of worrying about where their next meal will come from and whether they will have a roof over their heads that night, they have to suffer the terrible cramps, headaches and blood loss associated with menstruation. It can be quite taxing for an individual already undergoing many hardships to also be denied their basic right as a female human being.

WITM believes that menstrual products including pads, undergarments and more are extremely important for females to be able to have a clean and risk-free menstrual cycle. This group acknowledges that despite the gravity of this situation, these products are found in shockingly low amounts at places that homeless women and trans women could visit. Due to this ongoing issue, WITM has worked tirelessly with several organizations, including Seva Food Bank, Nannocare, Joni and Intimina, to provide these underprivileged individuals with the basic necessities for a safe menstrual cycle. Recognizing their efforts, the Ted Rogers Students' Society designated WITM's Menstruation Mission as the Best New Initiative of 2019. Moreover, in this year, this team has raised over $980 in donations with plans to triple this amount by the end of the project year. This amazing accomplishment was achieved despite the hardships that COVID-19 has brought on in terms of effective fundraising, holding events, and being able to reach out to individuals face to face. WITM took this challenge and overcame it through holding a holiday raffle in collaboration with Ryerson Women in Leadership- in which $592.53 were raised. This is just one of many initiatives brought on by this team, to ensure that underprivileged women, now more than ever, still receive safe and clean menstrual products.

The sad reality, however, is that organizations such as WITM, require much more support and fundraising to be able to help all homeless and trans women of Ontario. These women are being denied their basic human rights and are put at risk of many infections and maladies. It is interesting that one can receive water, toilet paper and soap for free in bathrooms and around the province. Yet something as important and trivial as menstrual products are treated as luxuries one must pay for. WITM counteracts this notion and strives to ensure women of these two discriminated parties can stay safe and not worry about something that they should already be receiving. In order to donate to this organization, visit https://gf.me/u/y2v7ph. Every dollar goes toward supplying menstrual products to a homeless or trans women who deserves more than they are currently receiving.

Ashley Meelu, BSc (University of Waterloo) is an undergraduate student with a background in Biomedical Sciences. She is an independent Article Writer associated with the Antarctic Institute of Canada.

The Plight of Those Experiencing Homelessness in the COVID-19 pandemic

Ashley Meelu, Jasrita Singh, Austin Mardon

In a certain suburban community, during a morning stroll, you may cross paths with a man who appears to be hauling his belongings on a wagon. He can be seen grabbing pizza at the local store with a smile and a sunny attitude. Nonetheless, behind that buoyant exterior, is a man worried about how to stay sheltered in light of the dangerous pandemic raging across the world.

Those experiencing homelessness ordinarily have a different normal than most others. Instead of lounging in their rooms, they hang out in public areas such as parks or coffee shops. Instead of playing video games, they walk around and converse with people. In a country where residents can chat up anyone from a Tim Hortons line-up to a bus stop, COVID-19 has taken away any desire for social interaction. However, those experiencing homelessness may find it difficult to adhere to regulations put in place by the government- such as quarantining or maintaining social distance. Where can someone that is experiencing homelessness quarantine? Even most homeless shelters can only guarantee a room for one night on a first come first serve basis.

Members of this marginalized population are now hard-pressed to risk their safety for a public place to rest or eat. At this point, there is risk of infection in those areas as many people might still be practising unhygienic methods. Thus, those experiencing homelessness are always at risk of exposure. However, some have no other choice than to risk it because they need a break. Moreover, in light of the pandemic, many restaurants and food chains have closed up and only agree to serve food through drive-throughs. This intensely limits or completely ceases the options that the homeless population have for sources of food. Essentially, the homeless populations of Canada have no safe haven from the virus. Due to this, they are at more risk than anyone else to transmit the virus. Unfortunately, they have no steady or sufficient sources to receive free protective wear or hand sanitizer.

The added stress of the pandemic and the increased hardship faced by those experiencing homelessness, certainly correlates with mental health issues. Due to this, it is extremely likely that the homeless population might find themselves having more run-ins with the law and with each other. The toll being taken on them mentally is unimaginably high, but they have no method to relieve that stress. It can also push more citizens to drug use in a method to provide an avenue of escape from real-world troubles. Thus, the criminalization of the homeless populations has increased since COVID-19 broke out. The pandemic is taking a toll on every aspect of their lives and it often becomes hard to keep a positive outlook on life during such times.

The sad truth of this new world is that the same regulations cannot apply to every individual. It becomes extremely difficult to try to fend for oneself daily and at the same time, try to follow the laws put in place during COVID-19. These laws are placed to protect every citizen yet those experiencing homelessness have little or no avenue to help them. They may be fined for breaking pandemic-specific laws, such as disobeying physical distancing orders- fines they cannot pay. In reality, this should shed some light on how one can aid others by being empathetic for the adversities that others may be facing. The fear being felt throughout the world today should push citizens to practise social precautions, if not to keep themselves safe, then in consideration for those who have no other option than to be in public places.

Ashley Meelu, BSc (University of Waterloo) is an undergraduate student with a background in Biomedical Sciences

Jasrita Singh, BHSc (McMaster University) is an undergraduate student with a background in Biochemistry, Biomedical Discovery and Commercialization.

Austin Albert Mardon, CM, FRSC (University of Alberta) is an adjunct professor in the Faculty of Medicine and Dentistry, an Order of Canada member, and Fellow of the Royal Society of Canada.

The SHESHAT Volume 3 is an anthology assembled under the supervision of Drs. Austin and Catherine Mardon. This work will be published and promoted to different platforms such as Lulu, Google Scholar, and PubMed under the Antarctic Institute of Canada (AIC) Charity.

The SHESHAT Volume 3 is not a double-blind peer-reviewed journal as most journals; however, all articles are peer-reviewed thoroughly by experienced premedical and graduate students, and Dr. Mardon. The articles accepted in this paper are authored by skilled Writers of the Antarctic Institute of Canada. This anthology serves to appreciate and showcase youth scholarly research in the fields of gender studies, COVID-19, and socioeconomic aspects of daily living to name a few.

The AIC would like to acknowledge the #RisingYouth Grant offered by the TakingItGlobal Charity to many Article Writers to fund their project and publication. There are no conflicts of interests to declare.

Special Thanks to the Editor, Daivat Bhavsar, and the Graphic Designer Ethan Saldana, for their relentless efforts in assembling SHESHAT Volume 3.